Sir Arthur Conan Doyle

Sherlock Holmes Investigates
Three Stories of Detection

Text adaptation by **Kenneth Brodey** and **Rebecca Raynes**

Activities by **Kenneth Brodey**

Illustrated by **Gianluca Garofalo**

U0106775

Editor: Rebecca Raynes
Design and art direction: Nadia Maestri
Computer graphics: Simona Corniola
Picture research: Laura Lagomarsino

© 2009   Black Cat

First edition: January 2009

Picture credits: The National Portrait Gallery, London: 4;
© Bettmann/Corbis: 7; The Granger Collection, New York:
8; © Ralph White/Corbis: 44; De Agostini Picture Library:
46, 71; © Hulton-Deutsch Collection/Corbis: 47; Mary Evans
Picture Library: 48; Mary Evans Picture library/Alamy: 87;
© 2005 TopFoto.co.uk: 89; UNITED ARTIST/Album: 119;
20TH CENTURY FOX/Album: 120.

We would be happy to receive your comments and
suggestions, and give you any other information concerning
our material.
http://publish.commercialpress.com.hk/blackcat/

The Publisher is certified by

 CISQCERT

in compliance with the UNI EN ISO 9001:2008
standards for the activities of 'Design, production,
distribution and sale of publishing products.'
(certificate no. 04.953)

ISBN 978 962 07 0451 2

Printed in China by Leo Paper

Contents

The text is recorded in full.

 These symbols indicate the beginning and end of the
passages linked to the listening activities.

Sir Arthur Conan Doyle by Henry L. Gates.

The Man Who Created Sherlock Holmes:
Sir Arthur Conan Doyle

Sherlock Holmes and Dr John Watson are two of the most famous and best loved characters in all of literature. But Sir Arthur Conan Doyle, their creator, was just as fascinating as his creations.

Arthur Conan Doyle was born on 22 May 1859 in Edinburgh, Scotland. His father was a civil servant [1] and when he was still a young man his father had to go and live in a mental hospital because he was an alcoholic and epileptic.

---

1. **civil servant** : somebody who works for the government.

His mother, on the other hand, was a strong woman who came from a distinguished military family. She filled Arthur with ideas of honour and chivalry [1] – two ideas that are present in all his writings and actions.

In 1876 Conan Doyle began his medical studies at Edinburgh University. Because he had very little money, he worked for a doctor called Joseph Bell. Joseph Bell amazed [2] his students because he could guess the jobs and lifestyle [3] of his patients by simply observing them carefully.

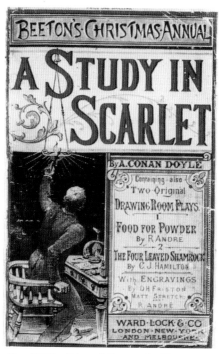

The 1887 edition of **Beeton's Christmas Annual**, which contained **A Study in Scarlet**.

Joseph Bell was Conan Doyle's principal model for Sherlock Holmes. When he finished his medical studies he started working as a doctor in Southsea, in the south England in 1882, but he was not successful. Very few patients came, so Conan Doyle had time to write. One of the books he wrote was a novel called *A Study in Scarlet*. This was the first Sherlock Holmes story.

1 **chivalry** : brave, polite, kind and unselfish behaviour.
2. **amazed** : greatly surprised.
3. **lifestyle** : how a person lives.

An 1891 edition of **The Strand Magazine**, which contained '**A Scandal in Bohemia**', the first Sherlock Holmes short story.

To make money, Conan Doyle sent a short story about Holmes to a popular monthly magazine called the *Strand*. Holmes soon became a big success.

The *Strand* immediately asked Conan Doyle for more stories about Sherlock Holmes, but from the beginning he had a strange relationship with his famous creation: he did not think his stories about Sherlock Holmes were serious and artistic enough. He wanted to write serious historical novels. So, from the beginning, Conan Doyle planned the death of Sherlock Holmes. When he told his mother about his plans to eliminate Sherlock Holmes forever she wrote to him, 'You won't! You can't! You mustn't!' But in 1893 Conan Doyle wrote a story called 'The Final Problem,' in which Holmes dies.

The reaction of readers was immediate. The *Strand* lost 20,000 readers, and people wrote thousands of letters to Conan Doyle asking him to bring Holmes back to life. Many people even insulted him. One woman wrote these eloquent words to him: 'You Brute.'

Finally, in 1901 Conan Doyle wrote a serialised Sherlock Holmes novel [1]

1. **serialised ... novel** : these were novels that appeared a chapter or a few chapters at a time in magazines.

Basil Rathbone as Sherlock Holmes in the 1939 film version of
The Hound of the Baskervilles.

called *The Hound of the Baskervilles*; the *Strand*'s circulation increased by thirty thousand copies. From then on Holmes appeared in the *Strand* until 1927, just three years before Conan Doyle's death.

But Conan Doyle's life was not just Sherlock Holmes. He was very active in public affairs. He spoke in favour of a Channel Tunnel, steel helmets [1] for soldiers and inflatable life jackets [2] for sailors. He also

1. **steel helmets :**   2. **inflatable life jackets :**

The 1917 photograph of **Frances Griffiths and the fairies**, which Conan Doyle
believed was real.

used his own analytic skills to solve crimes and to defend people
who were unjustly accused of crimes (see dossiers on pages 44 and
86). He was knighted in 1902 for his writing on the Boer War in
Africa and some said for his novel *The Hound of the Baskervilles*.

Conan Doyle also became very interested in spiritualism.
Spiritualism is the belief that it is possible to communicate with the
spirits of the dead. He also believed in the existence of fairies, [1] and
he said that some photographs of a little girl with fairies were real.
He even wrote a book in 1922 called *The Coming of the Fairies*.

1. fairies :

Of course, Conan Doyle, the creator of the most logical man in the world, Sherlock Holmes, was greatly ridiculed for these beliefs, but he did not seem to care.

He died in Crowborough, Sussex, England on 7 July 1930, one of the most famous and best loved men of his day.

The Sherlock Holmes Stories

The Sherlock Holmes stories were first published in the *Strand* magazine. The dates below say when these stories were first published as books.

A Study in Scarlet	(1890) This is the first Holmes novel and it appeared in a magazine called *Beeton's Christmas Annual*
The Sign of Four	(1890) a novel
The Adventures of Sherlock Holmes	(1892) Both 'The Blue Carbuncle' and 'A Case of Identity' appeared in this collection
The Memoirs of Sherlock Holmes	(1894) 'The Yellow Face' appeared in this volume
The Return of Sherlock Holmes	(1905) a collection of stories
The Hound of the Baskervilles	(1902) a novel
The Valley of Fear	(1915) a novel
His Last Bow	(1917) a collection of stories
The Case-Book of Sherlock Holmes	(1927) a collection of stories

Decide if each sentence is correct or incorrect. If it is correct, tick (✓) A; if it is incorrect, tick (✓) B.

	A	B
1 Conan Doyle's two most famous creations are Sherlock Holmes and Dr John Watson.	☐	☐
2 Conan Doyle's father was a banker.	☐	☐
3 Conan Doyle's father was the model for Sherlock Holmes.	☐	☐
4 Conan Doyle wanted to kill Sherlock Holmes because he did not think that his stories were serious and artistic enough.	☐	☐
5 When Conan Doyle published *The Hound of the Baskervilles* the *Strand*'s circulation increased by fifty thousand copies.	☐	☐
6 Spiritualism is the belief that you can solve any crime.	☐	☐

❷ The original Sherlock Holmes

Arthur Conan Doyle was a doctor. He described Sherlock Holmes as 'a scientific detective'. Sherlock Holmes's methods are very similar to the methods of a doctor. His 'detecting' work is like the 'diagnosing' work of a doctor. The 'clues' of the crime are like the 'symptoms' of a disease. In fact, Conan Doyle's model for Sherlock Holmes was a doctor called Joseph Bell. Below is an example of a conversation between Dr Joseph Bell — the original Sherlock Holmes — and one of his patients.

Dr Joseph Bell.

A Dr Bell : Well, my man, you've served in the army.

Patient : Aye, sir. (*'Aye' means 'Yes' in Scotland and various parts of northern England.*)

B Dr Bell : And you have left the army recently?

Patient : Yes, sir.

C Dr Bell : And you belonged to a Highland regiment? (*A Scottish regiment — 'the Highlands' are the mountainous region of northern and western Scotland.*)

Patient : Aye, sir.

D Dr Bell : And you were an officer?

Patient : Aye, sir.

E Dr Bell : And you were stationed at Barbados?
(*The most eastern island of the West Indies.*)

Patient : Aye, sir.

Now match the explanations given by Bell himself about how he was able to guess so many things about this man by simply observing him closely.

1 ☐ D He had an air of authority.

2 ☐ His disease was elephantiasis, a disease which occurs only in tropical countries, and which is also called 'Barbados Leg'.

3 ☐ The man was very respectful, but he did not remove his hat. In the army men do not remove their hat as a sign of respect; they salute — they raise their right hand to the side of their head.

4 ☐ He had a Scottish accent.

5 ☐ The man had not been a civilian long enough to have the habit of removing his hat as a sign of respect.

Read the text below and choose the correct word for each gap.

He is often sarcastic (**0**) ...C.... most definitely arrogant. However, he always does his best to help people and (**1**) .... a case. His laziness is incredible. He hates the (**2**) .... daily routine and desires only mental excitement. But when he has an interesting case, he can be (**3**) .... active. In his (**4**) .... time, he plays music. He lives at number 221B. What famous (**5**) .... creation are

we talking about? Mr Holmes of course. Well, maybe that was the right answer in the past, but now many people would answer Dr Gregory House of the extraordinarily popular medical show House. This TV show first appeared in 2004 and (**6**) .... many important awards. Finally, we should note (**7**) .... the name of the TV doctor 'House' is also a tribute to the great detective — 'Holmes' is pronounced just (**8**) .... 'homes'. So, a (**9**) .... doctor was the inspiration for a great literary detective, who was then the inspiration for a great TV doctor.

0	**A**	but	**B**	however	**C**	and	**D**	also
1	**A**	solve	**B**	repair	C	mend	D	answer
2	**A**	usual	**B**	natural	C	standard	D	common
3	**A**	much	**B**	too	C	greatly	D	quite
4	**A**	fun	**B**	game	C	free	D	relaxation
5	**A**	unreal	**B**	fake	C	fantastic	D	fictional
6	**A**	took	**B**	conquered	C	won	D	succeeded
7	**A**	that	**B**	which	C	how	D	why
8	**A**	as	**B**	similar	C	equal	D	like
9	**A**	real	**B**	correct	C	true	D	right

The Blue Carbuncle

Before you read

1 The characters

The characters of the first part of the story introduce themselves below. Match their introductions with the picture of each one.

A ☐ My name is Dr John Watson. I am a doctor. My wife takes good care of me, and always brushes my hat. I am a good observer, but, as my friend Holmes says, I do not reason with what I see.

B ☐ My name is Henry Baker. This is not a good time for me. I often drink, and my wife doesn't love me anymore. You can see this because my clothes are always dirty and my hat is never brushed.

C ☐ My name is Peterson. I am a commissionaire. For my job I wear a uniform that looks like a military uniform. I work at a hotel where I open the door for clients and take messages.

D ☐ My name is Sherlock Holmes. I am a scientific detective. I solve crimes by logical reasoning and with the help of scientific instruments like a forceps and magnifying glasses.

track 02
PET

2 Listening

You will hear Holmes and Watson talking about a man called Peterson — Peterson brought Holmes a goose and a hat. For each question, fill in the missing information in the numbered space.

An Instructive Hat

Watson arrives

Watson went to see Holmes (1) ....

How Peterson got the hat

Peterson was going home after a (2) .... It was very early in the (3) .... on Christmas Day. Some men attacked another man, and Peterson tried (4) .... In the end, the man who was attacked lost (5) ....

How Holmes got the hat

Peterson took the hat to Holmes on (6) .... Peterson knows that Holmes likes solving even (7) ....

Solving the mystery

Holmes thinks they can find the owner of the goose if they look carefully at (8) ....

3 Vocabulary

Write the words under the correct picture. Use a dictionary to help you.

magnifying glass   forceps   goose   walking stick   brim   wax

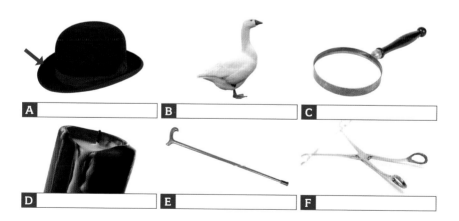

A	B	C

D	E	F

15

I visited my friend Sherlock Holmes two days after Christmas. When I arrived he was sitting on a sofa in front of the fire. Next to **track 02** the sofa was a wooden chair, and on the back of the chair was a dirty old hat. A magnifying glass and a forceps were on the chair, so the hat was probably part of one of Holmes's investigations.

'You're busy,' I said. 'Am I interrupting you?'

'No, not at all,' he replied, and indicated the hat. 'The problem is very simple, but it is still interesting and maybe even instructive.'

I sat down in an armchair and warmed my hands in front of the fire because it was very cold outside.

'I imagine,' I said, 'that this hat is connected with a terrible crime.'

'No, no. No crime,' said Sherlock Holmes, laughing. 'Do you know Peterson, the commissionaire?' [1]

'Yes.'

'This hat belongs to him.'

'It is his hat?'

'No, no. He found it. He does not know who the hat belongs to. Look at it carefully, and not as a dirty old hat, but as an

---

1. **commissionare** : a person in a uniform who stands outside public buildings or hotels to welcome people.

intellectual problem. It arrived here on Christmas morning together with a good fat goose. That goose is probably cooking at Peterson's house at this very moment.

'These are the facts. About four o'clock on Christmas morning Peterson was returning from a party along Tottenham Court Road. In front of him he saw a tall man carrying a white goose over his shoulder. Then he saw some men attack the tall man. One of the attackers knocked the man's hat off, [1] so the man lifted his walking stick to defend himself. But when he lifted the stick he broke the shop window behind him. Peterson ran to help the man, but when the man saw Peterson with his commissionaire uniform, he thought he was a policeman, and he ran away. The attackers ran away too and Peterson was there all alone with the hat and the goose.'

'Of course, Peterson then gave the goose back to its owner,' I said.

'No,' replied Holmes, 'that is the problem. It is true that "For Mrs Henry Baker" was written on a small card attached to the leg of the goose, and that the initials "H.B." are written inside the hat. But there are thousands of Bakers and hundreds of Henry Bakers in London and it is not easy to discover which Henry Baker it belongs to.'

'So what did Peterson do?'

'He brought both the goose and the hat to me on Christmas morning, because he knows that I am interested in even the smallest problems. I kept the goose until this morning and then I gave it to Peterson to cook for dinner. I am keeping the hat for the unknown gentleman.'

---

1. **knocked ... off** : pushed his hat off with force.

'Did the man who lost the goose put a notice in the newspaper?'
'No.'

'Then how can you discover who he is?' I asked.

'From his hat,' replied Holmes.

'You can't be serious! What can you learn from this dirty, old hat?'

'Here is my magnifying glass,' replied Holmes. 'You know my methods. Look at the hat and see what you can discover about the identity of the man.'

end

I took the hat and looked at it carefully. It was a very ordinary round black hat. It was very old and inside I could see the initials 'H.B.'. There was a hole in the brim for a hat-securer, 1 but there was no elastic. It was very dusty 2 and dirty in several places, but the owner had tried to cover these dirty places with black ink.

'I can see nothing,' I said, and gave the hat to Holmes.

'On the contrary, Watson, you can see everything, but you do not reason with what you see.'

'Then please tell me what you can deduce from this hat,' I said.

'Well,' said Holmes as he looked at the hat carefully, 'I can see that the man was very intellectual, and that three years ago he had quite a lot of money, but that recently he has had problems with money. He had foresight 3 in the past, but much less now, which means he has some problem. He probably drinks too much. This is probably the reason why his wife has stopped loving him.'

'My dear Holmes!'

---

1. **hat securer** : a band that holds the hat on the head.
2. **dusty** : covered with dust.
3. **foresight** : careful planning for the future.

'He has, however, kept some self-respect,' continued Holmes. 'He stays at home and goes out very little. He is totally unfit, [1] is middle-aged, has grey hair, which has been recently cut, and he uses hair cream. These are the main facts. Also, I don't think he has gas lighting in his house.'

'You're surely joking, Holmes.'

'Not at all. Don't you understand how I deduced these things?'

'I am certain that I am very stupid,' I replied, 'but I don't understand. For example, how did you deduce that this man was an intellectual?'

To answer me Holmes put the hat on his head. The hat was too big for him and covered his eyes.

'It's a question of volume,' said Holmes. 'If a man has such a big brain, he must have something in it.'

'How do you know he has less money now than in the past?'

'This kind of hat first came out three years ago. It is a hat of the very best quality. If this man had enough money to buy such an expensive hat three years ago, but he has not bought another hat since then, then it is clear that he has much less money now.'

'Well, that is clear enough, certainly. But how about the foresight?'

Sherlock Holmes laughed. 'Here,' he said, pointing at the hat-securer. 'Hat-securers are never sold with hats. This means that he ordered it, which is a certain sign of foresight. But since he has not put in some new elastic after the old one broke, this means that he has less foresight than before. But he has tried to hide some of the dirty marks on his hat with ink which means he has not completely lost his self-respect.'

'Your reasoning is certainly very good,' I said.

---

1. **unfit** : in bad physical condition because he does not exercise very much.

'That he is middle-aged, that his hair is grey, that his hair has been recently cut and that he uses hair cream can all be seen by looking closely at the inside of the hat. With the magnifying glass you can see the short pieces of grey hairs cut by a barber's scissors. They stick to the hat and there is the distinct odour of hair cream. Also, you will observe that the dust on the hat is the soft, brown dust you find in houses, not the hard, grey dust you find in the streets. This means that the hat is kept inside the house most of the time, and that he doesn't go out very often. Also you can see the marks of sweat [1] on the inside of the hat, which means he sweats a lot. A man who sweats so much can't be very fit.'

'But his wife — you said that she stopped loving him.'

'This hat has not been cleaned for weeks. When a man's wife lets him go out in such bad condition it means that she doesn't love him anymore.'

'But maybe he isn't married,' I said.

'No, he was bringing the goose to make peace with his wife. Do you remember the card on its leg?'

'You have an answer for everything. But how did you deduce that he doesn't have gas lighting in his house?'

'Well, if you saw only one or two marks of wax on a hat, it could be by chance. But I can see at least five on this hat, which means that this man must use candles very frequently.'

'Well, you're very clever,' I said, laughing, 'but since a crime has not been committed, all this seems to be a waste of time.'

---

1. **sweat** : the liquid that comes out of the skin when you are hot.

The text and **beyond**

1 Comprehension check
Read the first part of the story and answer the following questions.

1 When did Watson visit Holmes?

2 Was the hat connected with an important crime?

3 Why did Holmes think that the hat was interesting?

4 Who was Peterson?

5 Who attacked Henry Baker?

6 Why did the man carrying the goose run away when he saw Peterson?

7 Peterson knew that the man's name was Henry Baker, so why didn't he return the goose and the hat?

8 Why did Peterson bring Holmes the goose and the hat?

9 How was Holmes going to find out more about the identity of the man who lost the hat and goose?

2 A dirty hat
Match the clues with Sherlock Holmes's deductions about Mr Baker.

THE CLUES

1 There is dust on his hat, and it is the kind of dust you find in houses, not on the street.

2 There are some hairs sticking to the hat and there is the distinct smell of haircream.

3 There are at least five wax marks on his hat.

4 His hat is not brushed.

5 The hat is big.

6 He bought a hat-securer, but when it broke he did not replace it.

7 There are pieces of cut hair stuck to the inside of the hat.

8 There was the label on the leg of the goose which says 'For Mrs Henry Baker' and there were the initials 'H.B.' written on the inside of the hat.

9 His hair is greyish in colour.

10 This hat came out three years ago, and it is quite worn.

11 He has tried to cover up the marks with ink.

12 There are marks of sweat on the hat.

THE DEDUCTIONS

A ☐ He is intellectual.

B ☐ He had enough money in the past, but doesn't now.

C ☐ He had foresight, but he has less now.

D ☐ He has some self-respect.

E ☐ He is middle-aged.

F ☐ He has recently had his hair cut.

G ☐ He uses hair-cream.

H ☐ He doesn't go out very often.

I ☐ He is not very fit.

J ☐ His wife has stopped loving him.

K ☐ He is married.

L ☐ He does not have gas lighting in his house.

T: GRADE 5

● **Speaking: celebrations**
Find a picture about Christmas, if possible from your country. Tell the class about it using these questions to help you:

1 How is Christmas being celebrated in the picture?

2 What do people usually eat and drink at this celebration in your country?

3 What other countries celebrate Christmas?

④ The Baker Street irregulars

What are the Past Simple forms of the irregular verbs below in the box? All of them appear in the story.

> come put lie ~~run~~ see keep
> give lose say have/has

Now complete the following sentences with one of these verbs, creating either an affirmative (*I went*) or a negative sentence (*I didn't go*). These sentences should be true according to the story. You don't need all the verbs.

0 Holmes .didn't run........ to help the man. Peterson .ran................. to help the man.

1 Holmes .... that only important crimes are interesting.

2 Henry Baker .... a goose and a hat.

3 Watson .... many things when he looked at the hat.

4 Holmes .... the goose to Peterson.

5 Mr Henry Baker .... a lot of money.

6 Holmes .... the hat, but he .... the goose.

⑤ A chain of events

A **Below are the facts that we have learned so far. Number them in the right order.**

A ☐ The attackers run away.

B ☐ Peterson is alone with the hat and the goose.

C ☐1☐ Peterson is walking home from a party on Christmas morning.

D ☐ He decides to ask his friend Sherlock Holmes for help in finding the man who lost the hat and the goose.

E ☐ During the fight one of the attackers knocks off the man's hat.

F ☐ The attackers and the man who is being attacked see that Peterson is wearing a uniform and they think he is a policeman.

G ☐ Unfortunately, he drops the goose.

H ☐ He sees some men attacking another man who is carrying a goose.

I ☐ He wants to return the hat and the goose to the man but he doesn't know how to find him.

J ☐ The man runs away too.

B   Now rewrite the sentences as a paragraph, putting them in the past and using the connecting words below to put sentences 1 and 2, 4 and 5, 6 and 7, and 9 and 10 together.

so   but   when   so

> Peterson was walking home from a party on Christmas morning when he saw some men attacking another man who was carrying a goose.

6 Pronunciation

Below are 10 words from Part I. Find the word in the box that rhymes with each one. There are 9 words that you do not need to use.

hope   juice   close   taught   get   won   egg

last   taste   own   facts   meat   near

picks   nose   starred   scared   stop   wear

1 chair ....

2 goose ....

3 card ....

4 shop ....

5 alone ....

6 brought ....

7 leg ....

8 past ....

9 wax ....

10 sweat ....

PET 7 Sentence transformation

Here are some sentences from Part I of this story. For each question, complete the second sentence so that it means the same as the first, using no more than three words.

0 This hat has not been cleaned for weeks.
The last time this hat ....was cleaned was....... weeks ago.

1 The problem is very simple but it is still interesting.
This problem is interesting, ... it is very simple.

2 I warmed my hands in front of the fire because it was very cold.
It was .... I warmed my hands in front of the fire.

3 When the man saw Peterson with his commissionaire uniform, he thought he was a policeman.
The man thought Peterson was a policeman because .... a commissionaire uniform.

4 You can't be serious!
You .... joking!

5 It was a very ordinary hat.
There .... about the hat.

6 But maybe he isn't married.
But he .... married.

7 It is his hat.
The hat .... him.

8 Look at it carefully.
Take .... at it.

9 It was a very ordinary hat.
There .... about the hat.

Before you read

1 Listening

track 03

PET

Listen to the first section of Part II. For each question, put a tick (✓) in the correct box.

1 When Peterson rushed in he

   A ☐ smiled at Holmes.

   B ☐ looked incredibly shocked.

   C ☐ took off his coat.

2 What did Peterson's wife find in the goose's stomach?

   A ☐ a red necklace

   B ☐ an old shoe

   C ☐ a precious blue stone

3 Who did the stone belong to?

   A ☐ the Countess of Morcar

   B ☐ Peterson's wife

   C ☐ Sherlock Holmes

4 What was the blue stone?

   A ☐ a diamond

   B ☐ a blue carbuncle

   C ☐ a piece of glass

5 Holmes showed Watson and Peterson

   A ☐ an old book he read.

   B ☐ a newspaper article.

   C ☐ a letter from the Countess of Morcar.

6 Who discovered the robbery?

   A ☐ the Countess

   B ☐ Mr Horner

   C ☐ Mr Ryder

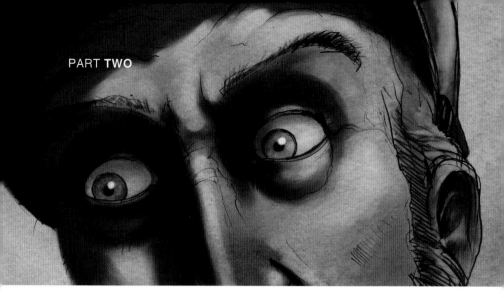

Sherlock Holmes had opened his mouth to reply, when the door opened and Peterson, the commissionaire, came running in. His face was red and he looked incredibly shocked.

'The goose, Mr Holmes! The goose, sir!' he cried.

'What? Has it come back to life and flown out of your kitchen window?' asked Holmes.

'Look here, sir! Look what my wife found in its stomach!' He showed us a shiny [1] blue stone in his hand.

'Peterson!' said Holmes. 'This is a treasure! Do you know what you have got?'

'A diamond, sir! A precious stone! It cuts glass like butter.'

'It is more than a precious stone. It's *the* precious stone.'

'Not the Countess of Morcar's blue carbuncle, [2] the one which was stolen?' I cried.

'Yes,' replied Holmes, 'I should know — I have seen the advertisement for it every day in the *Times* recently. The reward of £1,000 is certainly not even a twentieth of its value.'

---

1. **shiny** : bright, reflecting light.
2. **carbuncle** : a precious stone. Generally it is red but here it is blue, which makes it especially valuable.

'A thousand pounds!' cried Peterson. He sat down on a chair and looked at Holmes and me.

'It was lost at the Hotel Cosmopolitan, if I remember correctly,' I said.

'That's right,' replied Holmes. 'On 22nd December; just five days ago. John Horner, a workman was accused of stealing it. Here is the newspaper article about it:

Hotel Cosmopolitan Jewel Robbery

John Horner, a 26-year-old workman, has been arrested for stealing the famous blue carbuncle from the Countess of Morcar.

James Ryder, an employee at the hotel, said that he had sent Horner to the Countess's room on the day of the robbery to repair a part of the fireplace. Ryder said that he stayed with Horner for a few minutes, but then he had to leave. When Ryder returned he saw that Horner had gone and that somebody had forced open the Countess's writing desk. Ryder called the police and the police arrested Horner that same evening. Catherine Cusack, the Countess's servant, said that she heard Ryder call for help. She ran into the room and saw the same things that Ryder described to the police. In addition, the police discovered that Horner had been arrested for robbery in the past, but Horner says that in this case he is innocent. The case will be heard in court soon.

'Hm! So that's the police report,' said Holmes throwing the paper on a chair. 'You see, Watson, our little deductions about the hat have become much more important and less innocent.

Here is the stone: the stone came from the goose, and the goose came from Mr Henry Baker, the gentleman with the bad hat and all the other characteristics which we examined so carefully. Now we must discover Mr Baker's part in this mystery. To find him, the simplest thing is to put a notice in the evening newspaper.'

'What will you say?' I asked.

'Well,' said Holmes, ' "Found at the corner of Goodge Street a goose and a black hat. Mr Henry Baker can have them if he comes to 221b Baker Street at 6.30 this evening." '

Then Holmes sent Peterson to buy another goose to give to Baker if he came, and to put the notice in all the newspapers.

After Peterson left, Holmes held the carbuncle up against the light.

'It's beautiful,' he said. 'But it also attracts crime. Every good stone does. This stone isn't even twenty years old. It was found in Southern China and it has all the characteristics of a carbuncle except that it is blue and not red. It already has a violent history. There have been two murders, a suicide and various robberies because of this stone. Who could imagine that such a pretty stone could cause so much crime?'

'Do you think that Horner is innocent?' I asked.

'I don't know,' he replied.

'And do you think that Henry Baker has something to do with it?'

'I think that Henry Baker is absolutely innocent and didn't know what was inside the bird. But I will discover that with a little test if he sees the notice in the newspaper and comes here this evening.'

I left to work for the day. That evening, when I came back, I saw a tall man waiting outside Holmes's house. We entered together.

'Mr Henry Baker, I believe,' said Holmes when he saw us. 'Please sit by the fire and get warm. Ah, Watson, you have come at the right time. Is that your hat, Mr Baker?'

'Yes, sir, that is certainly my hat.'

Mr Baker had an enormous head and an intelligent face. His red nose and cheeks and shaking hand indicated his drinking habits. His clothes were old.

'We have kept your things,' said Holmes, 'but we had to eat the goose.'

'You ate it!' said our visitor with excitement.

'Yes, it was going to go bad, but I bought you another goose. It is over there, and I think it is just as good.'

'Oh, certainly, certainly!' answered Mr Baker happily.

'Of course,' said Holmes, 'we have the feathers, legs and stomach of your bird if you want them.'

The man laughed loudly. 'Perhaps I could keep them to remember my adventure, but, no, I don't need them. Thank you, but I will take this goose and go.'

'There is your hat, then, and there is your bird,' said Holmes. 'But could you tell me where you got your goose from? It was a splendid bird, and I would like to get another one like it.'

'Certainly, sir,' said Mr Baker, 'I got it at the Alpha pub near the Museum. You see, the owner of the pub, Mr Windigate, started a goose-club. Each week we gave him a few pence,[1] and then at Christmas we received a goose.'

---

1. **pence** : coins, the plural of penny.

After this Mr Henry Baker picked up his hat and goose, and left.

'So much for Mr Henry Baker,' said Holmes when Baker had gone, 'it is certain that he knows nothing about the stone.'

We decided to go immediately to the Alpha pub to investigate the goose. There we discovered that the goose had come from a stallholder [1] called Mr Breckinridge in Covent Garden Market. So, once again, Holmes and I put on our coats and walked to Covent Garden to talk to Mr Breckinridge.

'Remember,' said Holmes as we walked to Covent Garden, 'at one end of this chain of events we have a simple goose, but at the other end of the chain there is a man who will go to prison for seven years if we cannot show that he is innocent.'

We soon found Mr Breckinridge's stall, and Holmes asked him about his geese. I was surprised when Mr Breckinridge replied angrily to Holmes's questions.

'I have had enough. I am tired of people asking me "Where are the geese?" and "Who did you sell the geese to?" and "How much money do you want for the geese?" Enough!'

With a little bit of difficulty, Holmes finally got the information we needed: the geese had come from Mrs Oakshott, 117 Brixton Road. We were walking away when we heard shouting from Mr Breckinridge's stall. We turned round and saw a little man with a face like a rat in front of the stall.

'I've had enough of you and your geese! If you come here again, my dog will attack you!' shouted Mr Breckinridge at the little man. 'Bring Mrs Oakshott here, and I'll answer her. It's none of your business!'

'But one of them was mine,' he replied.

---

1. **stallholder** : person who owns a table in an outdoor market.

'Well, ask Mrs Oakshott about it, then,' shouted Breckinridge.

'She told me to ask you.'

'I don't care! Go away!'

The little man started walking away, and Holmes and I went after him. Holmes put his hand on the man's shoulder. The little man turned around and looked frightened. He said, 'Who are you? What do you want?'

'Excuse me,' said Holmes, 'but I heard you talking to the stallholder, and I think I can help you.'

'You? Who are you? How could you know anything about all this?'

'My name is Sherlock Holmes. It is my job to know what other people don't know.'

'But do you know anything about this?'

'Excuse me, I know everything about this. You are trying to find some geese which were sold by Mrs Oakshott, of Brixton Road, to a salesman called Breckinridge, who then sold them to Mr Windigate of the Alpha pub, who then gave one of them to a member of his goose-club called Mr Henry Baker.'

'You are the man I wanted to meet,' said the little man, 'I can't tell you how interested I am in this goose.'

'In that case we should go and talk somewhere warm,' said Holmes, 'but first, please tell me your name.'

The man hesitated. 'John Robinson.'

'No, no, your real name,' replied Holmes kindly.

The man became red in the face. 'My real name is James Ryder.'

'Exactly,' said Holmes, 'you work at the Hotel Cosmopolitan. Now, let's get into this cab and I will soon tell you everything you want to know.'

The man looked at us with frightened but hopeful eyes; he wasn't sure if something very good or very bad was going to happen to him.

We then returned to Holmes's house to discuss the matter in front of a warm fire.

'Here we are!' said Holmes happily, as we entered his room. 'Now do you want to know what happened to those geese?'

'Yes, sir,' replied Ryder.

'But you really want to know what happened to that goose — the white one with a black bar across its tail.'

Ryder trembled with emotion. 'Oh sir,' he cried, 'where did it go?'

'It came here.'

'Here?'

'Yes, and it was an incredible bird. I am not surprised that you want to find that goose. It laid an egg after it died — the brightest little blue egg that you have ever seen. I have it here.'

Our visitor stood up and then almost fell down. Holmes took out the blue carbuncle, and Ryder looked at it for a long time. He did not know if he should say it was his or not.

'The game is over, Ryder. I know almost exactly what happened. Because you worked at the Hotel Cosmopolitan you knew that the Countess of Morcar had the blue carbuncle in her room.'

'It was the Countess's servant, Catherine Cusack, who told me about it.'

'I see,' continued Holmes, 'so you and Catherine Cusack broke the grate ¹ in the Countess's room so that Horner had to come

---

1. **grate** : the metal frame in a fireplace.

and repair it. You knew that Horner had had a part in a robbery before so that he would be accused of this one. Then, when Horner had finished repairing the fireplace, you called the police and the unfortunate man was arrested. You then...'

Ryder threw himself onto the rug and held onto Holmes's knees, 'Please have mercy! Think of my father! Think of my mother! It would break their hearts. I have never done wrong before!'

'Get back into your chair!' said Holmes. 'It is easy to say that now, but you did not think of this poor Horner before.'

'I will go away, Mr Holmes, and without my declaration Horner will be free.'

'Hmm! We'll talk about that next,' said Holmes. 'And now tell us how the blue carbuncle came into the goose, and how the goose came into the open market. Tell us the truth because that is your only chance not to go to prison.'

Ryder moved his tongue over his dry lips and began his story.

'I will tell you exactly what happened. After I had the blue carbuncle I was terrified. I did not know where to go. I thought I saw the police everywhere. Finally I decided to go to my sister's. My sister married a man called Oakshott and lives on Brixton Road, where she keeps geese to sell at the market. When I arrived she asked me what was wrong. I told her that I was upset about the robbery at the hotel.

'I then went outside where the geese are, and smoked a pipe. I had a friend called Maudsley who had been in prison. He had told me how thieves [1] sell what they steal so I decided to go to him

---

1. **thieves** : (singular, thief) people who steal things, who take things that are not theirs.

with the blue carbuncle. However, I didn't know how I could carry the blue carbuncle to his house. Then I had the idea to force one of the geese to swallow [1] the stone. My sister had told me that I could have one of the geese for Christmas. So I caught

---

1. **swallow** : send something from your mouth to your stomach.

one of the geese — a big white one with a black bar across its tail, and forced open its beak [1] and pushed the stone in with my finger. The goose then swallowed the stone. Then I told my sister that I wanted my Christmas goose then. She thought it was a bit

1. beak :

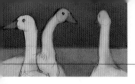
strange, but in the end she said I could have it.

'Unfortunately, while I was talking to my sister the goose escaped and went among the other geese. I caught it again, killed it and took it to my friend Maudsley. I told him the story. We then cut open the goose, but we couldn't find the stone! I ran back to my sister, and asked her if there were any other white geese with a bar across their tails. She said that there were two other ones, but she had sold them to a stallholder called Breckinridge in Covent Garden.

'I went to him, and he told me that he had sold them all. You heard him tonight. Now people will think I'm a thief, and I have not even touched the blue carbuncle. God help me!'

There was a moment of silence, and then Holmes got up and opened the door.

'Get out!' shouted Holmes.

'What sir? Oh thank you!' cried Ryder.

'No more words. Get out!'

And there were no more words. Ryder ran out of the room and out of the house.

'After all, Watson,' said Holmes, taking his pipe, 'if the police can't catch their own criminals, I don't have to do it for them. And this Ryder will never commit another crime again. He's too frightened. And this *is* the season of forgiveness. [1] Chance has given us an incredibly interesting little problem, and its solution is our reward. [2] And now, Doctor, let's begin another investigation in which a bird is also the most important part: our dinner.

---

1.   **forgiveness** : pardon, mercy.
2.   **reward** : prize, payment.

The text and **beyond**

PET **1** Comprehension check

Read the sentences below. For each question put a tick (✓) in the correct box.

1 Why did Peterson come to see Holmes?
   A ☐ because he had lost his Christmas goose
   B ☐ because his goose had died
   C ☐ because the goose had come to life again and flown out of the window
   D ☐ because he had found a precious stone in the stomach of the goose

2 How did Peterson know that it was a precious stone?
   A ☐ because it was shiny
   B ☐ because it cut glass like butter
   C ☐ because Peterson had read about it in the newspaper
   D ☐ because Holmes told him that it was a precious stone

3 Who is James Ryder?
   A ☐ He is the servant of the Countess of Morcar.
   B ☐ He is the hotel employee who sent John Horner up to the Countess's room.
   C ☐ He is a friend of the Countess of Morcar.
   D ☐ He is a commissionare, a friend of Sherlock Holmes.

4 Who is John Horner?
   A ☐ He is a hotel employee.
   B ☐ He's the man who repaired the fireplace in the Countess's room.
   C ☐ He is the Countess's servant.
   D ☐ He is a friend of James Ryder.

5 How does Holmes deduce that Mr Henry Baker had no part in the theft of the blue carbuncle?

A ☐ because he came to get his hat back

B ☐ because he dropped the goose and ran away from Peterson

C ☐ because he did not want the stomach of the first goose

D ☐ because he got the goose from the goose-club, and Mr Windigate did not work at the hotel

6 How did Holmes know that Ryder had stolen the blue carbuncle?

A ☐ because the goose salesman at Covent Garden told him

B ☐ because Ryder was looking for the goose, so he must have known that the blue carbuncle was inside it

C ☐ because Ryder had called the police to report the stolen blue carbuncle

D ☐ because Henry Baker had not stolen it, so Ryder must have stolen it

7 Why did Ryder force the goose to swallow the blue carbuncle?

A ☐ because he needed a safe way of carrying the blue carbuncle to his friend Maudsley

B ☐ because the goose was too thin

C ☐ because he was afraid that his sister would find it

D ☐ because he wanted to surprise Henry Baker

8 How did Breckinridge get the goose with the blue carbuncle?

A ☐ Ryder took the wrong goose. The goose with the blue carbuncle was sold by Ryder's sister to Mr Breckinridge.

B ☐ The goose flew to Covent Garden and Mr Breckinridge caught it there.

C ☐ Maudsley sold it to Mr Breckinridge by mistake.

D ☐ Mr Breckinridge went to Ryder's sister and asked her for the goose.

2 The whole chain of events

Number the following sentences in the right order to show how the blue carbuncle came into Sherlock Holmes's hands.

A ☐ Mrs Oakshott, James Ryder's sister, sells the goose to Mr Breckinridge.

B ☐ Mr Peterson takes the goose home, and gives it to his wife, who cuts it open to find, to her great surprise, the blue carbuncle!

C ☐ James Ryder takes the blue carbuncle to his sister's house and forces a goose to swallow it, but the goose escapes!

D ☐ While returning home with his goose, Mr Henry Baker is attacked by some men and he drops the goose and his hat, which are found by Mr Peterson.

E ☐ Mr Windigate of the Alpha Inn buys the goose from Mr Breckinridge for his Christmas goose-club.

F ☐ Mr Windigate of the Alpha Inn gives the goose to a member of his goose-club, Mr Henry Baker.

G ☐ Mr Peterson takes the goose and the hat to his friend Mr Sherlock Holmes — Holmes keeps the hat, but he gives the goose to Mr Peterson for his dinner.

H ☐ James Ryder, with the help of the Countess's maid, steals the blue carbuncle.

I ☐ Mr Peterson takes the blue carbuncle back to Holmes, who then begins his investigation of this incredible chain of events.

J ☐ The Countess of Morcar brings the blue carbuncle with her to the Hotel Cosmopolitan.

3 Put it in the past

Put all the verbs in the sentences above into the past and rewrite them as a paragraph using linking words.

The *Titanic* leaving Great Britain from the port of Southampton in the south of England.

Conan Doyle Defends
the Crew of the Titanic

Almost everybody knows about the *Titanic*, the luxury passenger ship which sank in the Atlantic Ocean on 15 April 1912. It has become a symbol of human arrogance and presumption.

The *Titanic* left the English port of Southampton on Wednesday 10 April at full speed. [1] It continued to travel at full speed even when it received four telegraph messages saying that there were icebergs in the area. The captain of the *Titanic* was certain that his crew would see the icebergs in time. In fact, at 11.40 p.m. on 14 April some

---

1. **full speed** : as fast as it could go.

members of the crew telephoned the captain and said, 'Iceberg right ahead!' The captain then gave orders to change the direction of the ship, but he was not in time.

The *Titanic* hit the iceberg and the collision caused a 300-foot cut in its hull. [1]

At about midnight the captain knew that the boat was going to sink, [2] and ordered the crew to begin preparing the lifeboats. Unfortunately, not all of the crew knew which lifeboats were theirs or which passengers had to get on their lifeboats. This caused a lot of confusion.

Finally, at 12.30 a.m. the captain gave the orders for the women and children to be put on the lifeboats, but many of them did not go: many did not realise that the *Titanic* was about to sink; others thought that another ship, the *Carpathia,* was nearby; others did not want to leave their husbands. Therefore, many lifeboats left half full, and although the sea was calm, very few of the lifeboats returned to help the passengers in the water after the *Titanic* had sunk.

At 2.30 a.m. the captain and his officers went down with the ship. Of the 2,200 passengers on board only about 700 survived.

The newspapers of the day wrote about the sinking of the *Titanic* in great detail. But according to the playwright [3] George Bernard Shaw the journalists did not write the truth. They wrote romantic lies. Conan Doyle, however, considered Shaw's attack on the journalists of the day as an attack on the crew of the *Titanic* itself and on the honour of Britain.

---

1. **hull** : principal body of a ship.
2. **sink** : (*sink, sank, sunk*) go under the water.
3. **playwright** : someone who writes plays (*Romeo and Juliet* is a play).

McElroy, the ship's purser, and Captain Smith on the *Titanic*.

They both wrote a series of long letters to the newspapers. Conan Doyle wrote in one of his letters that the band on ship played the religious song 'Nearer to God' to keep the passengers calm. Shaw responded that witnesses [1] reported that the band played happy, quick songs so that the passengers – especially the third-class passengers – did not realise that the *Titanic* was sinking until the lifeboats had gone.

Conan Doyle said that Captain Smith, the captain of the *Titanic*, was an honourable sailor who made one 'terrible mistake' and then tried to help save as many people as he could. Shaw replied that Captain Smith made no mistake. He knew perfectly well that icebergs are the only risk that is considered really deadly in his job and knowing it, he risked it and lost.

1. **witnesses** : the people who see something happen.

A newsboy selling newspapers breaking news of the sinking of the **Titanic**.

Conan Doyle thought it was terrible that Shaw criticised the officers of the *Titanic* because they had done their duty very well. Shaw said that the officers in lifeboats refused to save the people who were in the water after the ship sank because they were afraid.

Conan Doyle ended his last letter about the *Titanic* by saying that Shaw did not have the humanity that 'prevents a man from needlessly [1] hurting the feelings of others'. [2]

Shaw felt that when people were deeply moved [3] by a tragedy like the sinking of the *Titanic* 'they should speak the truth'.

The truth? Or respect for the feelings of people who have suffered? Who was right? Shaw or Conan Doyle?

1. **needlessly** : for no real reason.
2. **hurting ... others** : offending others.
3. **moved** : touched emotionally.

The *Titanic's* band – all of them died.

1 Comprehension check

Answer these questions.

1 What kind of ship was the *Titanic*?
2 Did the captain of the *Titanic* know that there were icebergs near his ship?
3 Why did many women not want to leave the *Titanic* and get on the lifeboats?
4 How many people were aboard the *Titanic*? How many survived?

2 Fill in the chart below.

What Arthur Conan Doyle thought	What George Bernard Shaw thought
The newspapers wrote good things about the crew of the *Titanic*.	1
2	The band on the ship played happy songs so that the poorer, third-class passengers would not know the ship was sinking, and therefore they would not try to get on the lifeboats reserved for the rich first-class passengers.
3	Captain Smith knew that there were icebergs in the area, but he took a chance and lost. Because of this many people died.
The crew of the *Titanic* did their duty well. You should not hurt other people's feelings for no reason.	4

You are one of the passengers that survived the tragedy of the *Titanic*. This is part of a letter you receive from a friend.

> I have just heard the terrible news! I am so glad that you are safe. Please tell me what happened.

Now write your letter in about 100 words.

 INTERNET PROJECT

The *Titanic*

Nearly one hundred years later, there is still great interest in the *Titanic*, and the Internet, of course, reflects this interest. To visit some of the more important sites go to www.blackcat-cideb.com or www.cideb.it. Insert the title or part of the title of the book into our search engine. Open the page for *Sherlock Holmes Investigates*.

Click on the Internet project link. Go down the page until you find the title of this book and click on the relevant link for this project. Then together with your partner prepare a short report about an interesting aspect of the *Titanic* such as:

▶ films made about the *Titanic*

▶ legends about it

▶ famous people aboard

▶ the discovery of the ship in 1985

Download and print any interesting pictures you find to use in your report.

A Case of Identity

Before you read

1 The characters

The characters of the first part of the story introduce themselves below. Match their introductions with the picture of each one.

A ☐ My name is Miss Sutherland. I am a typist. I normally wear glasses because I have bad eyesight.
I receive interest from the money which I inherited from my uncle. I am engaged to Mr Hosmer Angel.

B ☐ My name is Mr Windibank. I am a businessman and I work in the City. I don't want Miss Sutherland to go out with men. I don't like Miss Sutherland's father's friends. I often go to France on business because I buy wine there. I don't have a moustache, and I don't wear glasses.

C ☐ My name is Sherlock Holmes. I think normal, everyday life is very interesting and strange. If you observe people carefully, you will see that I am right. I told my good friend Watson that if we could fly over the houses of London, and look inside, we could see how interesting people's normal, everyday lives are.

D ☐ My name is Mrs Sutherland. I am Miss Sutherland's mother. After my husband died I married a younger man called Mr Windibank.

E ☐ My name is Mr Sutherland. I died a couple of years ago. Miss Sutherland is my daughter. My wife married soon after I died. Her new husband sold my plumbing business. I had a good business though.

F ☐ My name is Mr Hosmer Angel. I am very shy. I always whisper. I wear dark glasses. I have a moustache. I always type my letters to Miss Sutherland.

② What women wear

Match the names of the clothing with the pictures.

| dress | boots | gloves | hat | shoes | parasol |

T: GRADE 6

③ Speaking: clothes

Find a picture of someone wearing either old-fashioned or modern clothes which interest you. Tell the class about the clothes using these questions to help you:

1 What period are the clothes from?
2 Why do you like these clothes?
3 What clothes do you like wearing?

④ Physical characteristics

This story talks about what is distinctive (special) about a person. Answer the following questions.

A What is distinctive about you?

- Your physical appearance: hair colour, eye colour, the structure of your body, etc.;
  ....
  ....

- The way you talk;
  ....
  ....

- The way you write, (your handwriting);
  ....
  ....

- The way you dress?
  ....
  ....

  Describe what is distinctive about your partner.
  ....
  ....
  ....

B What could you change about yourself to hide your true identity?

C What do you think is distinctive about Mr Holmes? About Dr Watson?

D Describe what is distinctive about your favourite fictional character from books, film or TV.

5 Vocabulary

Match the pictures to the words. Use a dictionary to help you.

1 typewriter keys   **2** tonsils   **3** hansom   **4** false teeth

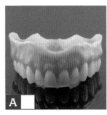  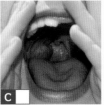 

A ☐   B ☐   C ☐   D ☐

track 04

PET

6 Listening

You will hear Holmes and Watson talking about what they think is interesting. For each question, put a tick (✓) in the correct box.

1 Holmes thinks that the most interesting things can be found in
   A ☐ police reports.
   B ☐ newspaper stories.
   C ☐ the regular lives of people.

2 Holmes knows all about Mr Dundas's situation because
   A ☐ he read about it in the newspaper.
   B ☐ it was part of one of his investigations.
   C ☐ he knows the people involved.

3 Holmes thinks important cases are not usually interesting because
   A ☐ you generally know why people committed these crimes.
   B ☐ you can never know why people commit these crimes.
   C ☐ their solutions are usually rather easy.

4 Holmes and Watson understand immediately that the big woman
   A ☐ can not decide to come to Holmes for help.
   B ☐ is not sure which doorbell is Holmes's.
   C ☐ is not sure whether Holmes will help her.

5 Holmes guesses that the woman
   A ☐ was treated badly by the man she loves.
   B ☐ does not understand the actions of the man she loves.
   C ☐ never again wants to see the man she loves.

'My good friend,' said Sherlock Holmes as we sat by the fire in his house at Baker Street, 'real life is much stranger than anything we could invent. If we could go out of that window, fly over London, and look into houses and see the peculiar things that are happening, we would discover things much more interesting than in books.'

track 04

'I'm not sure,' I answered. 'The stories in the newspapers are never very interesting. In fact, they are always very boring.'

'That is because,' said Holmes, 'newspapers always repeat the official reports of magistrates and police reports. You can be certain that there is nothing as unnatural and strange as what happens in everyday life.'

'I know,' I replied, smiling, 'that your cases are always very interesting, but let's look at today's newspaper.'

I picked up the newspaper. 'Now let's look at the first article I find. It's about a husband who was cruel to his wife. 'I don't have to read the article,' I said, 'but I am sure that the man had a girlfriend, that he drank too much and that he began to hit his wife.'

'You've chosen a bad example, Watson,' said Holmes, 'because I've worked on this case. The man, Mr Dundas, didn't

have a girlfriend and he didn't drink and he didn't hit her. Instead, at the end of every meal he took out his false teeth and threw them at his wife. You must admit that nobody could invent such a story!'

'Do you have any interesting cases now?' I asked.

'Well, I am working on ten or twelve cases, but none of them are interesting. They are important, you understand, without being interesting. I have found that unimportant matters are usually more interesting. If there is a big crime, the motive is generally obvious. So, they are generally not very interesting. But I think I will have an interesting case in a few minutes. That's a new client, if I'm not wrong.'

Holmes was standing at the window and looking down at the dark, grey London streets. There was a big woman standing opposite, looking up at our windows. She was moving her hands nervously. It was obvious that she wasn't sure whether to ring Holmes's doorbell or not. Then suddenly she ran across the road and rang it.

'I know those symptoms,' [1] said Holmes. 'When a woman hesitates like that on the pavement, it means that she has a love problem. She wants help, but she thinks that her problem is too delicate to communicate. But when a woman does not hesitate and rings the doorbell hard, it means she was treated very badly. In this case, we can presume that there is a love problem but that this woman is confused and perplexed, and wants an explanation.'

As Holmes was speaking, the servant announced Miss Mary Sutherland. She was a big woman with a wide pleasant face. She

---

1. **symptoms** : signs that something are wrong.

was wearing a hat with a red feather, a black jacket, a dark brown dress and grey gloves. She also wore small, round gold earrings. Holmes asked her to sit down and then looked at her carefully.

He said, 'Isn't it difficult for you to type with such bad eyesight.'[1]

'It was at first,' Miss Sutherland replied, 'but now I can type without looking at the keys.' Then she looked surprised and frightened when she understood that Holmes already knew so much about her.

'How do you know that?' asked Miss Sutherland.

'It is my business to know things,' said Holmes laughing. 'If I could not see these things, why would people come to me? In any case, I can see the marks of the glasses on your nose and the double line on your wrist where a typist presses against the table.'

'I have come here,' she said, 'because I want to know where Mr Hosmer Angel has gone.'

'Why did you come here in such a hurry?' asked Holmes.

Once again Miss Sutherland looked very surprised. Holmes then explained that her boots were not the same and that they were not completely buttoned.

'Yes, I did hurry out of the house because I was angry with Mr Windibank, that is, my father. He didn't want to ask the police about Mr Angel. He said that nothing bad had happened. This made me angry so I came here to see you.'

---

1. **eyesight** : if you have bad eyesight, you cannot see very well.

'Your father?' said Holmes. 'He must be your stepfather [1] because his surname is different from yours.'

'Yes, my stepfather. I call him father, although that seems strange to me. You see, he is only five years older than me.'

'And is your mother alive?' asked Holmes.

'Oh, yes, mother is alive and well,' answered Miss Sutherland, 'but I was not happy when she married Mr Windibank so soon after father died. Also, Mr Windibank is fifteen years younger than mother. Father was a plumber [2] and had a good business, and when he died mother continued the business. But when she married Mr Windibank, he made her sell it. They got 4,700 pounds for it but that was much less than its real value.'

I thought Holmes would not be interested in these unimportant facts. Instead he was listening with great concentration.

'Do you live on the money from the business?' asked Holmes.

'Oh no,' replied Miss Sutherland, 'I inherited some money from my uncle. I cannot touch it, but with the interest, I receive one hundred pounds a year.'

'That should be enough for you to live quite comfortably,' said Holmes. 'I'm sure you travel a little and enjoy spending your money.'

'Oh no, Mr Holmes. While I am living at home I don't want to cause them expense, so I give that money to mother, and I live on the money I make typing,' she replied.

'I understand. Now, can you tell us about Mr Hosmer Angel?' asked Holmes.

---

1. **stepfather** : if your mother has a husband after your real father, her new husband is your stepfather.
2. **plumber** : person who repairs water pipes, sinks, toilets, etc.

Miss Sutherland blushed [1] and said, 'I met him at the plumbers' ball. [2] They used to send tickets to my father when he was alive, and after he died they sent them to my mother. But Mr Windibank didn't want us to go. He never wanted us to go anywhere. He used to get angry every time I wanted to go out. He said that my father's friends were not good enough for us. But the day of the ball, Mr Windibank went to France on business, so mother and I went to the ball, and it was there I met Mr Angel.'

'I suppose,' said Holmes, 'that Mr Windibank was very angry with you when he discovered that you had gone to the ball.'

'No, not very,' replied Miss Sutherland, 'I remember he laughed and said that it was impossible to stop a woman when she really wanted something.'

'And did you see Mr Angel after the ball?' asked Holmes.

'Yes, he came to the house the next day, but after that father came home and so he couldn't come to the house. Father didn't want anybody to come to the house. So Mr Angel said, "We should wait until your father goes to France before we see each other. While he's at home we can write to each other every day."'

'Were you engaged to [3] Mr Angel at this time?' asked Holmes.

'Oh yes, Mr Holmes. We were engaged after the first walk that we took. Mr Angel worked in an office in Leadenhall Street.'

'Which office?'

'That's the worst part. I don't know.'

'Where did he live then?'

---

1. **blushed** : her face became red because she was embarrassed.
2. **ball** : party where you dance.
3. **engaged to** : if you are engaged to someone, you have promised to marry him/her.

'Above his offices.'

'Then where did you send your letters?'

'To the Leadenhall Street Post Office, where he went to get them. He said to me, "The other workers in my office will laugh at me if they see I am receiving letters from a lady."'

'I told him that I could type my letters, like he did his. But he said, "A typed letter comes from an impersonal machine and not from you." This shows how fond he was of me, Mr Holmes, and the nice little things he thought of.'

'Yes,' said Holmes. 'I have always said that the little things are the most important. Can you remember any other little things about Mr Angel?'

'He was a very shy man, Mr Holmes. He always wanted to walk with me in the evening instead of during the day. He was very gentlemanly. Even his voice was gentle. He told me that he had had a bad infection of the tonsils when he was a child, so he had a weak throat and had to talk very quietly. He always wore elegant clothes. His eyes were weak, just like mine, and he wore dark glasses because the sun hurt his eyes.'

'Well, what happened when Mr Windibank went back to France?' asked Holmes.

'Mr Angel came to the house and said that we should get married before father returned. He was very serious and said, "Put your hand on the Bible and promise me that you will always love me." Mother agreed with him. Mother liked him from the beginning, and liked him even more than I did. When they started talking about me and Hosmer getting married that week, I asked them if I should ask father first. They said no. But I did not want to do anything in secret, so I wrote to father at his

office in France to inform him. But the letter came back to me on the morning of the wedding.'

'It arrived after he left then?'

'Yes, sir, he had started back to England just before the letter arrived in France.'

'Ha! That was unfortunate. Your wedding was planned then for the Friday of that week. Was it going to be in church?'

'Yes, sir, but very quietly. On the day of the wedding Hosmer came in a hansom to take mother and me to the church. But since there were two of us, mother and I went in the hansom, and Hosmer took a cab. [1] We got to the church first, and when the cab arrived, we waited for him to get out, but he didn't. The cabman said that he could not understand what had happened to him. That was last Friday and I haven't seen him since.'

'I think that you have been very badly treated,' said Holmes.

'Oh no, sir! Hosmer was too good and kind to leave me like that. All morning before the wedding he said to me many times, "If anything happens to me, you must always love me. You must wait for me. I will return to you." I thought this very strange to say on the day of our wedding, but his disappearance explains it.'

'It certainly does,' said Holmes. 'In your opinion, did he know that he was in danger?'

'Yes, I think so.'

'But do you know what the danger was?'

'No, I don't.'

'One more question. How did your mother react?'

'She was angry and told me that I should never speak about him again.'

---

1. **cab** : another type of carriage.

'And your father? Did you tell him?'

'Yes, he said, "Something terrible has happened to Hosmer, but I think he will return." I agree with my father. Why would Hosmer leave me? Hosmer did not borrow money from me, and I never put the money which I had inherited in his name. So he did not take my money and leave. But what happened? And why hasn't he written? I can't sleep at night because I'm so upset.'

Then she pulled out a handkerchief, and began to cry.

'I will try to solve your problem,' said Holmes, 'but don't think about it anymore. Try and forget about Mr Angel.'

'Do you think that I will ever see him again?'

'No, I'm afraid not.'

'Then what has happened to him?'

'Leave that with me. Now, I need some of Mr Angel's letters, a good description of him, and also the address of your father's offices.'

'I never had Mr Angel's address,' said Miss Sutherland, 'but here is Mr Windibank's address. He works for a wine importer. Here is the advertisement with a description of Hosmer that I put in the newspaper the *Chronicle* and here are four letters from him.'

'Thank you,' said Holmes. 'Remember my advice. Consider the incident closed and do not allow it to affect your life.'

'You are very kind, Mr Holmes,' said Miss Sutherland as she was leaving, 'but I will always wait for Hosmer Angel to come back.'

The text and **beyond**

PET ① Comprehension check

Decide if each sentence is correct or incorrect. If it is correct, tick (✓) A; if it is incorrect, tick (✓) B.

	A	B
1 Holmes thinks that invented stories are much stranger than real life.	☐	☐
2 The man in the newspaper article drank, hit his wife and had a girlfriend.	☐	☐
3 Holmes does not think that big crimes are very interesting.	☐	☐
4 Miss Sutherland's real father is dead.	☐	☐
5 Mr Windibank is Miss Sutherland's stepfather.	☐	☐
6 Miss Sutherland receives a hundred pounds a year as interest.	☐	☐
7 Miss Sutherland is a typist.	☐	☐
8 Mr Windibank always went to the plumbers' ball.	☐	☐
9 Mr Windibank told Miss Sutherland to go to the plumbers' ball.	☐	☐
10 Mr Windibank was in France when Miss Sutherland went to the plumbers' ball.	☐	☐
11 Hosmer Angel wrote his letters to Miss Sutherland by hand because he did not like machines.	☐	☐
12 Miss Sutherland did not want to tell her stepfather that she was going to marry.	☐	☐
13 Miss Sutherland will always wait for Hosmer to return.	☐	☐

2 Word square

Find the words in the five different categories. Then find these same words in the word square below.

Written things

1 printed pages held together between two covers: _ oo _

2 a detailed written description: _ _ po _ _

3 a piece of writing in a newspaper or magazine: _ r _ _ c _ _

Love and marriage

4 the ceremony when you get married: _ e _ d _ _ _

5 If you are ....., you have agreed to marry someone: e _ g _ g _ _

Seeing things

6 the ability to see: _ _ _ si _ _ _

7 these help you see or protect your eyes: _ l _ _ _ _ _

The world of work

8 somebody who uses a typewriter: _ yp _ _ _

9 a place where a person or people do some business activity: _ ff _ _ _

10 Somebody who installs and repairs water pipes: _ _ um _ _ _

11 somebody who does work: _ _ rk _ _

12 somebody who brings in products from other countries: _ m _ ort _ _

Family

13 your male parent: _ at _ _ _

14 your female parent: _ ot _ _ _

15 your mother's husband who is not your real father: _ _ _ p _ at _ _ _

W	Z	P	L	U	M	B	E	R	X	C	K	O	T	O
O	E	P	L	E	A	S	E	D	O	N	T	F	T	Y
R	R	D	B	O	O	K	C	G	M	A	E	F	E	S
K	M	A	D	E	T	F	A	T	H	E	R	I	Y	O
E	O	U	S	I	N	M	A	B	U	S	I	C	E	L
R	T	A	R	A	N	G	V	U	L	G	A	E	S	S
M	H	I	L	G	A	G	A	S	R	I	S	O	I	T
I	E	K	L	A	L	A	S	G	R	E	D	N	G	E
M	R	P	O	R	P	A	M	U	E	S	P	R	H	P
P	J	O	H	T	J	C	S	H	I	D	L	O	T	F
O	M	I	K	I	A	I	S	S	M	Y	M	O	R	A
R	O	F	A	C	C	S	T	E	E	P	F	A	T	T
T	G	L	A	L	K	E	E	Y	E	S	I	T	E	H
E	B	I	G	E	H	A	T	S	O	O	N	B	N	E
R	M	O	M	I	S	T	Y	P	I	S	T	T	A	R

'He must be your stepfather...'

Miss Sutherland says her father's name is Mr Windibank. Holmes concludes that he is not her real father, but her stepfather. He says, *'He must be your stepfather because his surname is different from yours.'* If your friend orders a big bottle of mineral water you can say, *'You must be very thirsty because you have ordered a big bottle of mineral water.'*

3 **Match the evidence from the story in column A with the conclusions in column B. Then write sentences as in the example above.**

A

A 1 (According to Watson!) The story is in the newspaper.

B ☐ The woman in the street hesitates to ring Holmes's doorbell.

C ☐ Miss Sutherland has marks on both sides of her nose.

D ☐ Mr Angel wears glasses.

E ☐ (According to Miss Sutherland!) Mr Angel did not want typed letters from Miss Sutherland.

F ☐ Miss Sutherland says that she will always wait for Hosmer Angel to return.

B

1 be very common and boring   4 have a love problem

2 wear glasses   5 fond of her

3 have weak eyes   6 be in love with him

A According to Watson, the story must be very common
and boring because it is in the newspaper.

B ....

C ....

D ....

E ....

F ....

69

4 Truth is stranger than fiction

A In English there is the expression 'Truth is stranger than fiction.' Is there a similar expression in your language? Do you agree with this?

B In Part I Sherlock Holmes says, 'Real life is much stranger than anything we could invent.'
This is similar in meaning to 'Truth is stranger than fiction' but it means that strange, peculiar things are normal and commonplace. Do you agree? Do strange things happen to you?

5 True or false?

Below are three stories. Which ones do you think were invented, and which ones do you think are real? Discuss with your partner.

A THE REPENTANT THIEF
A man from Portsmouth in the south of England lost his wallet. He had left it in a phone booth. A week later, a man sent this letter and the wallet to the office of a local newspaper:
'I thought about it for a long time. I am sorry that I took the wallet.'
Inside the wallet, along with the original contents, the thief had added 45 pounds!

B THE YOUNG ADVENTURER
After school a 12-year-old boy went to the airport and got on an aeroplane. Nobody stopped him. Nobody asked him for a ticket. He sat down in the first-class section. The boy was discovered during the flight to Jamaica. The boy's mother said, 'I don't think my son understands what he did. He thinks that it was fun!'

C THE BAD HUSBAND WAS A BAD MURDERER
In Buenos Aires, Argentina, a man wanted to kill his wife. They lived on the eighth floor of a building. The man threw his wife out of the window, but she did not fall to the ground because she got tangled in some electric wires. The man saw that his wife was stuck in the wires so he decided to jump on her to kill her. Unfortunately for him, he missed his wife and fell to the ground and was killed. His wife was able to climb onto the balcony of a neighbour's flat.

Now look at page 126 and see if you were right.

Before you read

1 **What men wear**
Match the names of the clothing and facial hair with the picture.

trousers   beard   shirt   moustache   shoes   bowtie
jacket   top hat   waistcoat

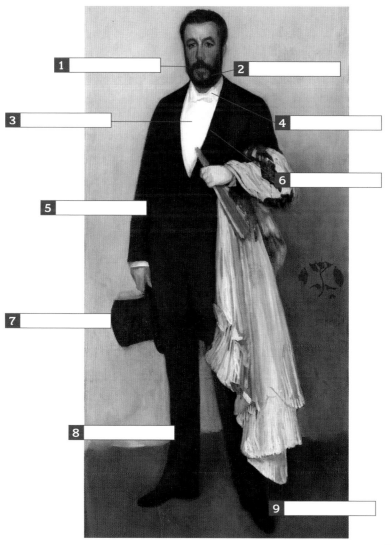

track 05

PET

2 Listening

Listen to the beginning of Part II of the story. Choose the correct picture and put a tick (✓) in the box below it.

1 Hosmer Angel's height is about

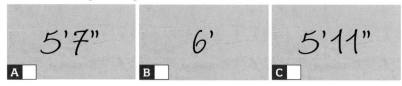

5' 7" | A ☐

6' | B ☐

5' 11" | C ☐

2 How did Hosmer sign his letters?

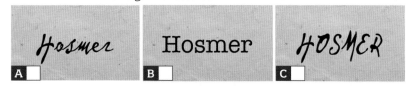

Hosmer | A ☐

Hosmer | B ☐

HOSMER | C ☐

3 What was Holmes doing when Watson returned the next day?

A ☐   B ☐   C ☐

4 What colour were Mr Windibank's eyes?

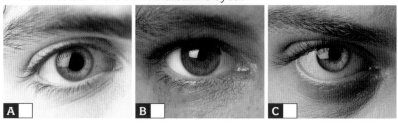

A ☐   B ☐   C ☐

5 Which letter of Mr Windibank's typewriter is not right?

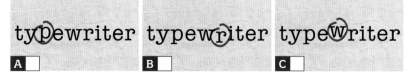

typewriter | A ☐

typewriter | B ☐

typewriter | C ☐

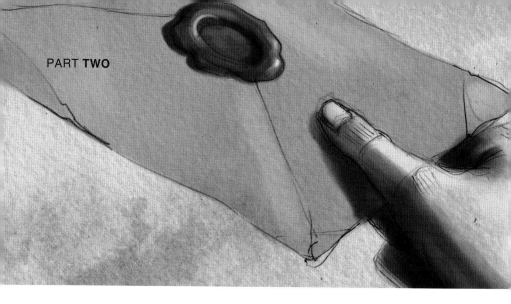

After she had left, I asked Holmes about the case.

'The young woman is more interesting than her little problem, which is not very difficult or unusual. Would you mind reading me the description of Hosmer Angel?'

track 05

I then read it to Holmes:

> *Missing, a gentleman called Hosmer Angel. About 5ft. 7in. tall. Well-built with black hair, a little bald in the centre; a black moustache; wears dark glasses; has a speech defect* [1]. *Wearing a black coat, black waistcoat, grey trousers and brown boots. Please contact Miss Sutherland...*

'That's enough,' said Holmes. 'Now look at these letters which Hosmer wrote to her. What do you see?'

'They are typed,' I said.

'Not only that, but the signature is typed too. This suggests a lot — in fact, we can call it conclusive.'

'Of what?'

'My dear Watson, can't you see how important this fact is to the case?'

---

1. **defect** : imperfection; Hosmer doesn't speak clearly.

'No, I can't,' I replied, 'unless Hosmer didn't sign his letters because he didn't want to be legally responsible for what he promised.'

'No, that was not the point,' said Holmes, 'but now I'll write two letters which will solve this mystery. One of the letters is to Mr Windibank's company and the other one will be to Mr Windibank himself to ask him to come here to meet us tomorrow evening at six o'clock.'

A few minutes before six the next day I returned to Baker Street. When I walked in, Holmes was doing chemistry experiments.

'Well, have you solved it?' I said as I walked into the room.

'Yes, it was the bisulphate of baryta.'

'No, no! Miss Sutherland's mystery!' I cried.

'Oh, that! I thought you were asking me about the chemistry experiment. There was never any mystery in the matter. The only problem is that the man who was involved did not do anything illegal, so he can't be punished.'

'Who was Hosmer Angel, and why did he leave Miss Sutherland?'

But Holmes did not have time to answer me, because exactly at that moment we heard someone walking towards Holmes's room and knock at the door.

'This is the girl's stepfather. He wrote to me to say that he was coming,' said Holmes. 'Come in!'

The man who entered the room was well-built, without a moustache and he looked at us with a pair of penetrating grey eyes.

'Good evening, Mr James Windibank,' said Holmes. 'I believe this is the typed letter that you wrote to me to say that you were coming here!'

'Yes, sir. I am sorry that Miss Sutherland has disturbed you with this little problem because I don't like other people knowing about our family misfortune. I didn't want her to come and see you. Anyway, I don't think that you will ever find this Hosmer Angel.'

'On the contrary,' said Holmes quietly, 'I am almost certain that I will find him.'

Mr Windibank was very surprised and dropped his gloves. 'I am happy to hear that,' he said.

'It is a curious thing,' remarked Holmes, 'that a typewriter is just as distinctive as a man's handwriting. For example, in this letter of yours, I can see that this part of the 'r' has a slight defect. There are also fourteen other characteristics of your typewriter.'

'We write all the letters in the office with this typewriter,' said Mr Windibank.

'And now,' continued Holmes, 'I will show you what is really very interesting. In fact, I am thinking about writing a book on the typewriter and its relation to crime. I have four letters here from Mr Angel. They are all typed and they all have a defect on the "r".'

Mr Windibank jumped out of his chair. 'I cannot waste [1] time over this ridiculous talk. If you can catch the man, catch him, and let me know when you have caught him.'

'Certainly,' said Holmes, walking over to the door and locking it. 'I am letting you know now that I have caught him.'

'What! Where?' shouted Mr Windibank becoming white, and looking around like a rat in a trap.

'You can't get away, Mr Windibank,' said Holmes. 'This case was really very easy. Now sit down and we can talk about it.'

---

1. **waste** : not use well, use for stupid things.

Mr Windibank fell back into the chair. 'I did not do anything illegal,' he cried.

'I am afraid that you are right. But, Mr Windibank, it was a cruel and selfish[1] trick. Now, let's look at what happened,' said Holmes.

Then Holmes sat down and began to talk. Mr Windibank sat with his head in his hands.

'The man marries a woman older than himself for her money. He can also use the money of the daughter while the daughter lives with him and the mother. The daughter has a lot of money so it is important not to lose it. But the daughter is friendly and affectionate, so it is clear that she will soon find a husband. At first this man tells the daughter that she cannot go out, but this will not solve the problem forever. Then one day the daughter says that she wants to go to a ball. What does the clever stepfather do then? With the help of the wife, he disguises himself.[2] He wears dark glasses, and puts on a fake[3] moustache. Then he changes his voice and speaks very quietly. He is even more certain that his plan will work because the girl has bad eyesight. Then he appears at the ball and keeps away other lovers by becoming the girl's lover himself.'

'It was just a joke at first,' said Mr Windibank. 'We didn't think that the girl would fall in love.'

'Yes, that's probably true,' continued Holmes. 'But the girl really fell in love, and you decided to take the situation to the extreme.

1. **selfish** : a selfish person only thinks about himself and never thinks about other people.
2. **disguises himself** : changes his appearance so that other people will not recognise him.
3. **fake** : false.

You began to see her often, and the mother said that she liked him very much. You asked Miss Sutherland to marry you so that she would never again think about other men. But it was difficult for you to pretend[1] to go to France every time Miss Sutherland had to see Mr Angel. You had to end the situation dramatically. In some way, you had to keep Miss Sutherland from thinking about other men in the future. Therefore, you made her promise on the Bible, and you told her that something could happen on the very morning of the wedding. You took her to the church, but obviously you could not marry her. You disappeared by using the old trick of entering one door of a cab and walking out the other. I think this is what happened, Mr Windibank!'

'Yes, maybe that is true,' replied Mr Windibank, 'but I did not do anything illegal, and now you are breaking the law because you will not let me leave this room.'

'You are right. You did not do anything illegal,' said Holmes as he unlocked and opened the door, 'but you really should be punished, and I would like to do it.'

Then Holmes picked up a riding-whip[2], but Mr Windibank ran out the door and out of the house.

'Now, he certainly is cold-blooded,' said Holmes laughing. 'That man will continue doing worse and worse crimes until he does something really bad and is hanged. In any case, this case had some interesting points.'

'I cannot completely follow your reasoning[3] in this case,' I said.

'Well, it was clear from the first, that Mr Hosmer Angel had a

---

1. **pretend** : make people believe something which is not true.
2. **riding-whip** : you hit horses with this to make them go faster.
3. **your reasoning** : how you reasoned, how you thought.

very good reason for his actions, and that the only man who could really profit from the situation was the stepfather: he wanted to keep the hundred pounds a year. Then Mr Windibank and Mr Hosmer Angel were never together, and the dark glasses, the quiet voice and the moustache all suggested a disguise. The final point was the typed signature. This made me think that the handwriting of the man must be very familiar to Miss Sutherland, and that if she saw even a little of it, she would recognise it.'

'And how did you confirm these ideas?' I asked.

'First I wrote to Mr Windibank's office. In the letter I described Mr Angel after I had eliminated everything that could be a disguise, like the glasses, the moustache and the voice, and I asked them if they had an employee like that. They wrote back to me and said that I had described Mr James Windibank. Then I wrote to Mr Windibank to invite him here, and as I expected he typed his reply to me. Then I compared his letter with the letters of Mr Angel. That's all!'

'And Miss Sutherland?' I asked.

'If I tell her, she will not believe me,' replied Holmes. 'Maybe you remember this Persian saying, "It is dangerous to take a tiger cub [1] from its mother, and it is dangerous to take a delusion [2] from a woman."'

---

1. **tiger cub** : baby tiger.
2. **delusion** : something you believe in which is not true.

The text and **beyond**

1 Comprehension check

Match the phrases in columns A and B to make complete sentences.
There are five phrases in column B that you do not need to use.

A

1 ☐ Hosmer didn't sign his letters
2 ☐ Holmes was doing a chemistry experiment
3 ☐ Holmes did not have Mr Windibank arrested
4 ☐ Mr Windibank was shocked when Holmes said that he had found Mr Angel
5 ☐ Holmes studied typewriters
6 ☐ Mr Windibank did not want Miss Sutherland to marry
7 ☐ Mr Windibank had Mr Angel disappear
8 ☐ Holmes did not tell Miss Sutherland what really happened to Hosmer

B

A because he had already solved Miss Sutherland's problem.
B because he didn't want to make her angry.
C because they helped him to catch criminals.
D because he had not done anything against the law.
E because he preferred to hit him with a riding-whip.
F because Mr Angel did not really exist.
G because they present many interesting mechanical problems.
H because he would lose money.
I because he was not really in love with her.
J because he didn't want to make her sad.
K because he didn't want Miss Sutherland to recognise his writing.
L because he didn't want to be legally responsible for his actions.
M because he could not go to France all the time.

2 **Missing: a gentleman called Hosmer Angel**

Read the notes about Hosmer Angel.

> *Missing, a gentleman called Hosmer Angel. About 5ft. 7in. tall. Well-built with black hair, a little bald in the centre; a black moustache; wears dark glasses; has a speech defect [1]. Wearing a black coat, black waistcoat, grey trousers and brown boots. Please contact Miss Sutherland...*

Now look at the following three pictures and write similar descriptions using the additional information.

A **Effie Munro:**   about 5ft. 5in./ in the case of 'The Yellow Face'/ contact Sherlock Holmes

C **Professor Moriarty:** criminal / arostocratic / English accent / the Napoleon of crime / contact Sherlock Holmes

B **Reginald Musgrave:** old friend of Sherlock Holmes / aristocratic accent / contact Sherlock Holmes

PET ❸ Sentence transformation

Here are some sentences about the characters on the previous page. For each question, complete the second sentence so that it means the same as the first using no more than three words.

0   Professor Moriarty doesn't speak with a foreign accent.
    Professor Moriarty hasn't ..*got a foreign*........ accent.

1   Professor Moriarty is the only man who has ever defeated Holmes.
    No .... defeated Holmes apart from Professor Moriarty.

2   If no one contacts Sherlock Holmes, he won't be able to find these missing people.
    Sherlock Holmes won't be able to find these missing people unless .... him.

3   Reginald Musgrave has got more hair than Professor Moriarty.
    Professor Moriarty has got .... Reginald Musgrave.

4   Effie Munro hasn't worn pink for a long time.
    It's a long time since Effie Munro .... pink.

5   Professor Moriarty is known as the Napoleon of crime.
    People .... Professor Moriarty the Napoleon of crime.

T: GRADE 5

❹ Speaking: transport

Find a picture of a means of transport. Tell the class about it using the following questions to help you.

1   Do you use this means of transport?

2   How did people travel in the past?

3   How will they travel in the future?

5 Conan Doyle and the case of the missing cousin

Because Conan Doyle was very famous as the creator of Sherlock Holmes, people often wrote to him for help. In 1907, a Scottish woman wrote and asked him to find her missing cousin.

Dear Mr Doyle,

I need your help. My cousin disappeared a week ago. I think something terrible has happened to him. Maybe he was kidnapped! He went to London, and he stayed in a big hotel. The evening he disappeared, he went to a music-hall show. He returned to the hotel about ten o'clock. But nobody saw him after that. The man who was staying in the next room said, 'I heard noises in his room during the night.'

Please help me!

Conan Doyle said, 'I tried to see the problem through the eyes of Mr Holmes.' He telephoned the bank of the missing cousin. The missing cousin had taken all of his money — £40 — from the bank before he came to London.

After an hour, Conan Doyle sent the woman a message:

Your cousin is in Scotland.
Look for him in Glasgow or Edinburgh.

How did he know this? Here is how he reasoned, but you must put his conclusions on page 85 in the right places!

The missing cousin wanted to disappear because he had taken out all his money before he disappeared.

A ⌐6⌐ Therefore, he was not kidnapped.

The man in the next room was wrong. There are many noises in a big hotel.

B ☐ Therefore, ...

The missing cousin left during the night. But there is a night porter in all hotels. It is impossible to leave a hotel without the night porter seeing you after the door is shut. The door is shut at midnight.

C ☐ Therefore, ...

The missing cousin left with his bag. No one noticed him leaving.

D ☐ Therefore, ...

The missing cousin wanted to hide, but he did not want to hide in London. If he had wanted to hide in London, he would not have left the hotel.

E ☐ Therefore, ...

If the missing cousin took a train to a small town, people would notice him.

F ☐ Therefore, ...

I looked at the train timetables and discovered that the only trains going to big cities were the trains going to Edinburgh and Glasgow.

G ☐ Therefore, ...

THE CONCLUSIONS:

1 It is probable that he left when many of the hotel guests were returning after the theatre at around eleven or eleven-thirty. This is when most shows end. After eleven-thirty very few people return to the hotel, and the missing cousin would have been noticed with his bag.

2 He would go to a big city, where most people get off the train and he would not be noticed or observed.

3 I am sure that he went to Edinburgh or Glasgow.

4 The missing cousin left before midnight.

5 He wanted to catch a train to go to some other place.

6 He was not kidnapped.

7 He probably heard noises from another room, and not noises from the missing cousin's room.

Conan Doyle Defends
the Underdog [1]

In Arthur Conan Doyle's books, his heroes always defend the underdog. You can see three examples in the stories in this book. Sherlock Holmes was a sort of modern knight, [2] and the poor and the weak knew that he would help them. But Conan Doyle did not only write about defending the underdog, he did it in real life.

One of Conan Doyle's battles was in favour of the Africans of the Congo, present-day Zaire. In 1885 King Leopold II of Belgium said that all 'empty' land in the Congo was his. 'Empty' meant that no Europeans lived there. Of course, there were many Africans who lived there, but for King Leopold these people did not have any importance. The only thing of importance was rubber, [3] especially for the new automobile industry. Rubber, at this time, [4] came from the trees of the African jungle, and it was very important then for industry. The Belgians forced the Africans to collect rubber. Often, if the Africans did not collect enough rubber, the Belgians cut off one of their hands. Sometimes, the Belgians kidnapped all the women and children of an African village. The men of the village could not have their families back until they brought the Belgians a certain

---

1. **Underdog** : someone who always loses because he is in a bad or disadvantageous position.

2. **knight :**

3. **rubber** : strong elastic substance used to make tyres, balls, raincoats, etc.

4. **rubber ... time** : Before World War II, all rubber came from the juice of rubber trees. During the war synthetic rubber was developed, and today most of the rubber used is synthetic.

A 1909 photograph showing the rubber industry in the **Belgian Congo**.

amount of rubber. A British diplomat, Roger Casement, began to protest against the treatment of the Africans. Conan Doyle joined Casement's protest, and in 1909 wrote a book, *The Crime of the Congo*. Conan Doyle also travelled around Britain and spoke about the crimes of the Congo. He also wrote letters to important people like the President of the United States, Theodore Roosevelt.

Later, after World War One, in 1918, Roger Casement was convicted of treason [1] because of his fight for Irish Independence. Even though

---

1. **treason** : the crime of betraying your country.

Conan Doyle was not for Irish independence, he fought for Casement not to be hanged but didn't succeed.

Another of Conan Doyle's famous battles for the underdog concerned a young solicitor called George Edalji. George's father was Indian and his mother was English. George's father was a vicar in a small English town, and he began to receive anonymous threatening letters. [1] At the same time, a lot of horses were attacked and physically injured. The police accused George of both the anonymous letters and the horse-maiming. [2] He was condemned to seven years in prison.

Once in prison, George wrote an article defending himself in a magazine. Conan Doyle read this article and believed what George wrote. He began to investigate the anonymous letters and the horse-maiming. He discovered that George was a very hard-working, calm young man, who never drank alcohol and who was never cruel. He also discovered that George couldn't see very well. In fact, George could not see anything that was more than six metres away. This was an important fact because the horse-maiming was committed at night, and the criminal had to cross many railway lines and go around many obstacles in the dark. This was almost impossible for George.

After learning these things, Conan Doyle wrote many newspaper articles defending George. Finally, the Government looked at George's situation again. They decided that he was not responsible for the horse-maiming, but that he was still responsible for writing the anonymous letters. So, after three years, George left prison, but he received no monetary compensation for his time in prison because of the letter writing.

1. **threatening letters** : letters in which the person said he was going to do something bad to George's father.
2. **horse-maiming** : attacking and physically hurting horses.

UNSOLVED..

The heartless slaughter of animals at night, a wrongful jail sentence, and a murder threat to the Home Secretary were combined in

The Cattle Maiming Outrages

BY

BERNARD O'DONNELL

Continuing his series of famous crimes which have defied detection

George Edalji, who was wrongfully sentenced for the cattle maiming.

GEORGE EDALJI
Wrongly sentenced for the cattle maiming outrage, and later pardoned. The culprit has never been traced.

Conan Doyle then found the person who had done the horse-maiming and written the letters. The criminal had told someone that he had done some horse-maiming, that he was an expert butcher and that he had written anonymous letters. In addition, the criminal's handwriting was the same as the handwriting of the letters which George's father had received. Finally, the horse-maiming continued when George was in prison, but it stopped when the criminal himself was away from the area. With this evidence, [1] Conan Doyle

---

1. **evidence** : the information which you use to make a decision, to prove something, etc.

was sure that the Government would say that George was not guilty of writing the anonymous letters. But the Government did nothing, even with the best evidence possible. So, George received no money for his time spent in prison as an innocent man. Sherlock Holmes would not have been very happy in the real world.

1 Comprehension check
Answer the following questions.

1 Why is Sherlock Holmes a kind of modern knight?
2 How was Conan Doyle like Holmes?
3 What did King Leopold mean when he described the Congo as empty?
4 Why had rubber become so valuable?
5 When did natural rubber stop being so important?
6 Why did it become less important?
7 Why did Conan Doyle and Roger Casement think King Leopold's actions in the Congo were criminal?
8 Why was George Edalji sent to prison for seven years?
9 What was George Edalji like?
10 What was the evidence Conan Doyle found to help George?

2 Discussion
There is an expression in English that says 'Crime does not pay'. However, to judge from Conan Doyle's experience we could say 'Helping the underdog does not pay'.
With your partner discuss briefly the two questions below. Then report your ideas to the class. Which opinions are most common?

1 Was Conan Doyle right to help the underdog?
2 Nowadays, is it useless to help the underdog?

The Yellow Face

Before you read

1 **Reading pictures**
Look at the picture on page 99.

1 What can you see in the picture?
2 What is he/she looking at?
3 What does he/she see?

2 **Listening**

track 06

PET

You will hear about Holmes, physical exercise and a new case. For each question, put a tick (✓) in the correct box.

1 Holmes generally did physical exercise
   A ☐ in the spring when the weather was good.
   B ☐ when his cases required it.
   C ☐ when Watson was able to persuade him.

2 Holmes became a little angry because
   A ☐ he didn't like taking walks.
   B ☐ he thought he had lost an interesting case.
   C ☐ the man had left his pipe.

3 Holmes examined the pipe and concluded that
   A ☐ it was important to the man.
   B ☐ it was very expensive.
   C ☐ it would soon break.

4 The man left Holmes's house
   A ☐ to look for Holmes outside.
   B ☐ to return to his wife.
   C ☐ to wait for Holmes.

5 The man's problem concerned
   A ☐ his wife.
   B ☐ two strangers.
   C ☐ his pipe.

Although he was one of the best boxers I have ever seen, Sherlock Holmes considered physical exercise as a waste of energy. He rarely took any exercise unless it was for professional reasons. Then he would never become tired. But one spring day I persuaded him to go for a walk with me in the park. We walked for two hours, and it was almost five when we returned to Baker Street.

'Excuse me, sir,' said our servant, as we entered, 'there was a man waiting for you. He was a very nervous gentleman. He kept walking all around the room saying, "Isn't Mr Holmes going to return?" Finally after half an hour he left saying he was going to wait outside in the fresh air.'

'You see,' Holmes said to me, 'I needed a case, and now I have lost this one because we went for a walk in the park. It's very annoying. Look! That's not your pipe on the table. Well, that man must have a very big problem because he left his pipe. It's obvious that he likes this particular pipe very much.'

'How do you know that he likes it very much?' I asked.

'Well,' explained Holmes, 'I think this pipe costs around seven-and-

sixpence. [1] Now, look it has been repaired twice with silver [2] bands that probably cost more than the pipe itself. So, this man must like his pipe very much if he prefers to repair it instead of buying a new one with the same money. But I think I can hear him coming up the stairs, so we will have something more interesting than his pipe to study...

A moment later a tall young man walked into the room without knocking. He was well dressed in a dark grey suit and was carrying a brown hat in his hand. He looked about thirty.

'Excuse me,' said the man, 'I should have knocked, but I am very worried and I need help.'

The man then sat down on a chair.

'I see you have not slept for a night or two. Now, how can I help you?'

'I need your advice, sir. I don't know what to do. It's very embarrassing. It isn't nice to talk about the behaviour of one's wife to two strangers.'

'My dear Mr Grant Munro...' began Holmes.

Our visitor jumped from his chair. 'What!' he cried. 'You know my name?'

'If you don't want people to know who you are,' said Holmes smiling, 'then you should not write your name on the inside of your hat, or else you should turn the inside of your hat away from the person who you are talking to.

'Anyway, my friend and I have heard many strange secrets in this room, and we have had the good luck to help many people. Please tell us the facts of your case.'

1. **seven-and-sixpence** : an amount of money, just over a third of a pound.
2. **silver** : precious white, shiny metal used to make jewellery, coins, knives, forks, etc.

Our visitor put his hand to his forehead as if it was very difficult for him to speak. I could see that he was a reserved, proud man, who wasn't used to talking to people about his private life.

'The facts are these, Mr Holmes,' he said. 'I have been married for three years, and my wife and I were very happy until last Monday. Suddenly she has become like a stranger to me. I want to know why. But, Mr Holmes, I am sure that my wife loves me. I know it. I feel it.'

'Please let me have the facts, Mr Munro,' said Holmes, with some impatience.

end

'Effie, my wife, was a young widow, only twenty-five years old when I met her. Her name then was Mrs Hebron. She went to America when she was very young and lived in the town of Atlanta, [1] where she married a man called Hebron, who was a lawyer. They had one child, but there was a yellow fever [2] epidemic there, and both her husband and child died of it. I have seen his death certificate. After this tragedy, she decided to leave America, and come back to England to live with her aunt.

'I should also mention that her husband left her a large amount of money. This money was invested, and she can live very well with the income [3] from it. She met me after six months in England. We fell in love with each other, and we married a few weeks afterwards.

---

1. **Atlanta** : city in the state of Georgia in the southern United States.
2. **yellow fever** : an often fatal disease caused by a virus transmitted by mosquitoes (see page 105).
3. **income** : the money you receive regularly for your work or, from an investment.

'I am a hop merchant, [1] and I, too, have a good income. We rented a nice house in the country near Norbury. There is a pub and two houses near our house, and a single cottage across the field in front of our house. Until last Monday my wife and I lived very happily there.

'There is one more thing I should tell you. When we married, my wife put all her money in my name. I did not think this was a good idea, but she insisted. Well, about six weeks ago she came and asked me for some.

'"Jack," she said, "when you took my money you said that if I ever wanted some, I should just ask you."

'"Certainly," I said, "it's your money. How much do you want?"

'"One hundred pounds," she said.

'"What for?" I asked, very surprised by the large amount because I thought that she wanted a new dress or something like that.

'"Oh," she said playfully, [2] "you said that you were only my banker, and bankers never ask questions, you know."

'"If you really want it, you will have the money," I said.

'"Oh yes, I really want it."

'"And you won't tell me what you want it for?" I asked.

'"Maybe one day but not now," she answered.

'I was not happy about this because this was the first time that there was a secret between us. I gave her the money and forgot about it. It may have nothing to do with what happened afterwards, but I thought that I should mention it.

---

1. **hop merchant** : person who buys and sells hops — plants whose flowers are used to make beer.
2. **playfully** : in a friendly and jokey manner.

'Anyway, I told you there is a cottage near our house. It has been empty for about eight months. Well, I like walking past that cottage, and last Monday, as I walked past it I saw an empty van [1] going away from it, and furniture in front of it. Someone was finally going to live there.

'I was looking at the cottage, wondering what type of people had come to live there, when suddenly I saw a face watching me from an upstairs window. There was something strange about the face, Mr Holmes, that frightened me. I was not very near, but there was something unnatural and inhuman about the face. It was yellow and rigid. I walked closer to the house, but the face suddenly disappeared. I was so disturbed that I decided to find out more about these people. I went to the door and knocked. A tall thin woman answered the door. I told her that I was her neighbour, and asked her if she needed any help.

'"If we need any help, we'll call you," she said and shut the door in my face.

'All evening I couldn't stop thinking about what had happened. I did not tell my wife about the strange face and the rude woman, but I did tell her that people were now living in the cottage. She didn't reply.

'That same night something strange happened. I usually sleep very deeply but that night I was woken up by a noise; it was my wife. She was dressed and was leaving the room. She looked very frightened and nervous. After she had gone out of the room, I looked at my watch. It was three o'clock. Why was my wife going

1.   **van** : large carriage for transporting things.

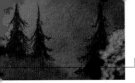

out at three in the morning? I waited for about twenty minutes, and then she returned.

'"Where have you been, Effie?" I asked as she entered. She moved back quickly with fright.

'"Are you awake, Jack?" she cried with a nervous laugh. 'I thought you were a deep sleeper!'

'"Where have you been?"' I asked again, more angrily.

'"I didn't feel well and felt that I had to go outside and get some fresh air. Now I feel much better."'

'While she was telling this story, she never looked at me and her voice was different. I knew she wasn't telling the truth. But what was my wife hiding from me?

'The next day I didn't go to work because I was so worried about my wife. At breakfast we didn't talk and I knew from the way she was looking at me that she knew I hadn't believed her story. After breakfast I went out for a long walk and was returning home at one o'clock when I walked past the cottage. I stopped for a minute in front of it to look for that strange face. As I stood there, imagine my surprise, Mr Holmes, when the door suddenly opened and my wife walked out! When she saw me her face became white. She looked very frightened as she came towards me.

'"Oh, Jack!" she said, "I came here to see if our new neighbours needed anything. Why are you looking at me like that? Are you angry with me?"

'"So," I said, "this is where you went during the night?"

'"What do you mean?" she cried.

'"You came here. I am sure of it. Who are these people?"

'"I have not been here before."

'"I know you are not telling me the truth. I am going into that cottage to find out the truth!"

'"Please, don't go in, Jack," she cried. "I promise that I'll tell you everything some day, but if you go in now, you'll cause great sadness." Then she held me tightly,[1] and I tried to push her off.

'"Trust me,[2] Jack!" she cried. "You won't be sorry. If you come home with me, everything will be alright. If you force your way into that cottage, our marriage is finished."

'"I'll trust you," I said, "on one condition. You must promise that this mystery finishes now. You can keep your secret but you must promise never to come here again."

'She looked much happier. Then, as we started to leave, I looked up and there was that yellow face watching us out of the upper window. What connection could there be between that person and my wife?

'After that everything went well for two days, but on the third day I returned home early. My wife wasn't in the house. The servant told me that she had gone out for a walk. But I didn't believe this. I went upstairs and saw the servant. She was running across the field towards the cottage. So my wife was there! I was very angry and went quickly to the cottage. While I was going there, I met my wife and the servant, who were returning, but I did not stop to speak to them; I wanted to discover the secret in the house.

'I walked into the house and found no one, but upstairs I found a comfortable room, and in it there was a full-length photograph

1. **tightly** : closely, with force.
2. **Trust me** : believe me.

of my wife. I left the house with a heavy heart. When I returned home I told my wife that there could be no peace between us until she told me the truth. That was yesterday, Mr Holmes, and then I decided to come and see you for help.'

After hearing this strange story, Holmes sat silent for a few minutes, thinking. Then he said, 'Are you sure that the yellow face was a man's face?'

'Each time I saw it,' he replied, 'I saw it from a distance, so I am not sure.'

'When did your wife ask you for the money?' asked Holmes.

'Almost two months ago.'

'Have you ever seen a photograph of her first husband?'

'No, there was a fire in Atlanta after her husband's death, and all her papers were destroyed.'

'But she had a death certificate. Have you ever seen it?'

'Yes, she got a duplicate after the fire.'

'Have you ever met anyone who knew your wife in America?'

'No.'

'Has she ever talked about visiting America again?'

'No.'

'Has she ever received letters from there?'

'No.'

'Thank you,' concluded Holmes. 'Now, go back to Norbury, and when you see that those people have returned to the cottage, call us. It should be easy to solve this mystery.'

The text and **beyond**

1 Comprehension check

A Complete the questions with the correct question word: *why,
how, who, what, where.*
If you can't decide which question word is correct, look at
the answers to help you.

THE QUESTIONS

1 .W.h.y....... was Holmes angry with Watson when they returned
from their walk in the park?

2 .... did Holmes know that the man liked his pipe very
much?

3 .... did Holmes know the man's name, even though the
man did not introduce himself?

4 .... is Effie?

5 .... did Effie live in America?

6 .... was her American husband's job?

7 .... did Effie and Mr Munro meet?

8 .... is Mr Munro's job?

9 .... much money did Effie want?

10 .... did Mr Munro see in the upper window of the cottage?

11 .... did Mr Munro see walk out of the cottage on the day
he returned early from the City?

12 .... did Mr Munro find on the mantelpiece of the cottage?

13 .... hasn't Mr Munro ever seen a picture of Effie's first
husband?

14 ...., if there was a great fire in Atlanta, has Effie got her
husband's death certificate?

B   Now match the questions with the answers given below.

THE ANSWERS

A ☐ In England.
B ☐ He was a lawyer.
C ☐ Grant Munro's wife.
D ☐ Because she obtained a duplicate after the fire.
E ☐ Atlanta, Georgia.
F ☐ He is a hop merchant.
G ☐ A full-length photograph of his wife.
H ☐ Because it was destroyed in the great fire in Atlanta.
I ☐ Because it was a pipe which didn't cost very much, but the man had repaired it twice with expensive silver bands.
J ☐ Because it was written on the inside of his hat.
K ☐ A rigid yellow face.
L 1 Because Holmes had lost a client.
M ☐ One hundred pounds.
N ☐ Effie.

2 The international battle against yellow fever
Read the text and fill in the gaps with the words in the box. Use a dictionary if you need help.

> disease   vaccine   germ   just
> later   developed   confirmed   epidemic

The yellow fever **(1)** ..epidemic...... of 1878 killed about 20,000 people in the southern United States. **(2)** .... three years **(3)** .... a Cuban doctor called Carlos Finlay proposed the theory that yellow fever was a caused by a **(4)** .... transmitted by mosquitoes. In 1900 the American army doctor Walter Reed **(5)** .... Finlay's theory: this was the beginning of an effective battle against the **(6)** .... However, it was not until 1937 that the South African scientist Max Theiler **(7)** .... a vaccine against the disease. Sadly, today about 30,000 people die each year of yellow fever because they are too poor to buy the **(8)** ....

'Have you ever seen a photograph of her first husband?'

Notice that we use the Present Perfect with *ever* and *never* in these cases:

I have never met him. = In my entire life I have not met him.

Have you ever seen her? = In your entire life have you seen her?

Notice when you talk about habits and routine with the Present Simple you can also use *never* and *ever*, but the meaning is different.

Do you ever go to the cinema? = Do you sometimes go to the cinema?

I never go to the cinema. = I don't go to the cinema at all.

3 A With the cues given below write questions using *ever*, and give true answers. Then, for the questions with negative answers, write sentences using *never*.

0   Mr Munro/see a photograph of his wife's first husband?

   Has Mr Munro ever seen a photograph of his wife's first husband?
   No, he hasn't.
   He has never seen a photograph of his wife's first husband.

1   Mr Munro/mend his pipe?

2   Effie/be to the United States?

3   Effie/ask Mr Munro for money before?

4   Mr Munro/be inside the cottage?

5   Holmes/see the yellow face?

6   Mr Munro/see the death certificate of Effie's first husband?

7   Mr Munro/meet anyone who knew his wife in America?

8   Effie/be inside the cottage?

B **Now write 6 questions to ask your partner using *ever*. Ask and answer.**

Before you read

1 **Prediction**

How many different explanations can you think of for the yellow face in the window?

Discuss your ideas in pairs and then share them in class. How many explanations did you think of?

2 **Listening**

You will hear about the solution of the mystery. Decide if each sentence is correct or incorrect. If it is correct, put a tick (✓) in the box under A for YES. If it is not correct, put a tick (✓) in the box under B for NO.

		A	B
1	Holmes thinks that Effie's first husband is still alive.	☐	☐
2	He thinks that he has come to England to blackmail her.	☐	☐
3	He also thinks that her husband's friend has a terrible yellow face.	☐	☐
4	Mr Munro sends Holmes a message asking him to come to Norbury.	☐	☐
5	Mr Munro does not wish to enter the cottage with Holmes and Watson.	☐	☐
6	Effie tells Mr Munro that he must enter the cottage.	☐	☐
7	They go into a room and see a girl in a red dress.	☐	☐
8	Watson is frightened when he sees the yellow face.	☐	☐
9	Watson is shocked when he sees the girl without the mask.	☐	☐
10	Mr Munro is shocked.	☐	☐
11	Effie refuses to explain the situation to Mr Munro.	☐	☐

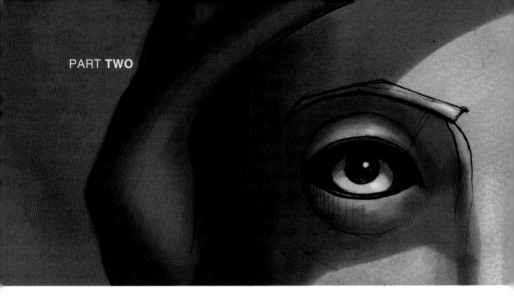

PART **TWO**

When Mr Grant Munro left, Holmes and I discussed the case.

'I am afraid that this is a case of blackmail,'[1] said Holmes.

'And who is the blackmailer?'[2] I asked.

'Well, it must be that person with the yellow face. I must say, Watson, there is something very interesting about that yellow face at the window, and I would not abandon this case for anything.'

'Have you got a theory?' I asked.

'Yes,' Holmes replied, 'I think her first husband is in the cottage. This is what I think happened: This woman was married in America. Her husband got a terrible disease.[3] That is why he has that horrible yellow face. She ran away from him at last, and came back to England, where she changed her name and started a new life. After three years of marriage, she felt safe again, but her first husband, or some unscrupulous woman attached to him, discovered where she lived. They wrote to her and told her to send them a hundred pounds, or they would tell her new

---

1. **blackmail** : asking for payment in return for not telling information, secrets, etc.
2. **blackmailer** : somebody who blackmails.
3. **disease** : illness caused by a germ (virus or bacteria).

husband everything. When her husband told her that someone was living in the cottage, she knew that they were her blackmailers. In the middle of the night, while her husband was sleeping, she decided to go to the cottage. That night she was not able to convince her blackmailers to leave her in peace so she returned the next day. That was when her husband saw her coming out of the house. She then promised her husband that she would not return, but she wanted her blackmailers to go away. She decided to go again, and this time she brought a photograph, which they probably asked her for. Fortunately for her, her maid warned her that her husband was coming, and she and her blackmailers left the house in time.

'Now we can do nothing except wait for Mr Munro to call us, and then we'll see if my theory is correct.'

We did not have to wait a long time. After tea we received a message from Mr Munro saying, 'There are people in the house. I have seen the yellow face again. I will do nothing until you arrive.'

That night Holmes and I took a train to Norbury. Mr Munro was waiting for us at the station. He was pale and agitated. 'They are in the house now, Mr Holmes,' he said, 'we'll go and discover the mystery.'

When we arrived there, Holmes asked Mr Munro if he was sure he wanted to go into the cottage. Mr Munro said he was sure and we went up to the door of the cottage. As we got near the door, a woman suddenly appeared. It was Effie.

'Please, Jack, don't!' she cried. 'Trust me!'

'I have trusted you too long, Effie!' he cried. 'Let go of me! My friends and I are going to solve this mystery.'

We went quickly up the stairs to the lighted room. In one corner there was a desk, and at that desk there appeared to be a

little girl. Her face was turned away from us when we entered the room, but we could see she was wearing a red dress and long white gloves. She turned around to us, and I gave a cry of surprise and horror. Her face was the strangest yellow colour and it had absolutely no expression.

A moment later the mystery was explained. Holmes, with a laugh, put his hand behind the ear of the little girl, and pulled off the mask, and there was a little black girl. She laughed, and I laughed too, but Grant Munro stood looking at her with his hand holding his throat.

'My God!' he cried, 'what does this mean?'

'I will tell you everything,' cried his wife. 'You have forced me, and now we must both accept the situation. My husband died at Atlanta. My child survived.'

'Your child?'

She pulled out a locket. [1]

'You have never seen this open.'

'I thought it didn't open.'

She opened the locket and inside was the picture of a very handsome and intelligent man; a man who was obviously of African descent.

'This is John Hebron, of Atlanta,' said Mrs Munro, 'and he was a very noble man. I cut myself off [2] from my race to marry him, but I was never sorry for a moment. Our only child looked like his people rather than mine. She is very dark, but she is my dear little girl.' When the little girl heard these words, she ran to her mother.

1. **locket :**

2. **cut myself off** : separated myself.

'I left her in America with a trusted servant,' Mrs Munro continued, 'because she wasn't very healthy, but I never wanted to abandon her. When I met you by chance and learned to love you, I was afraid to tell you about my child. I was afraid to lose you. I kept her existence a secret from you for three years, but finally I had to see my little girl. I sent the servant a hundred pounds, and told her to come to this cottage. I took every possible precaution so that there would not be talk about a little black girl. That is why she wore that yellow mask.

'You told me about her arrival in the cottage. I was so excited and that night I had to see her. That was the beginning of my troubles. And now, tonight, you know everything. What are you going to do about me and my child?'

Mr Grant Munro did not say anything for ten minutes, and his answer was one of which I love to think. He lifted the little child, kissed her, and, with the little girl in his arms, he gave his other hand to his wife.

'We can talk about it more comfortably at home,' he said. 'I am not a very good man, Effie, but I think that I am better than you thought.'

We all left the cottage together, and then Holmes and I returned to London.

We did not say another word about the case until late that night at Holmes's house in Baker Street, just before Holmes went to bed.

'Watson,' he said, 'if you should ever think that I am becoming too sure of myself, or that I am not working hard enough on a particular case, please whisper [1] "Norbury" in my ear, and I will be very thankful to you.'

---

1. **whisper** : say quietly so that no one else can hear.

The text and **beyond**

1 Comprehension check
Answer the following questions.

1 When did Holmes and Watson go to Norbury?
2 Why did Effie have to cut herself off from her race?
3 What did they first notice about the little girl at the desk?
4 How did Watson and the little girl react when Holmes pulled off the mask?
5 Who did the little girl look like?
6 Why did Effie leave her little girl in America?
7 How does Watson describe the man in the locket?
8 Why was Effie afraid to tell her husband about the little girl?
9 What was Grant Munro's reaction to the little girl before the explanation? And after the explanation?
10 What did Holmes say about his theory after he discovered the truth?

2 Getting the facts straight
In this story there are different interpretations of the facts. Holmes himself is wrong. Mr Munro doesn't understand what he sees, and Effie lies to maintain her secret. Below are three different explanations of seven parts of the story. Say which one is:

A The real explanation.
B What Mr Munro has seen or what Effie first told him.
C Sherlock Holmes's explanation.

1 ...... The husband died of yellow fever but the child survived.
...... After Effie's husband and child died, she returned to England to stay with an aunt.
...... Effie fled from her husband because he had a terrible disease.

2 ...... After three years of happy marriage, Effie wanted to see her daughter again.

...... After three years of happy marriage, Effie has some secret.

...... After three years of happy marriage Effie receives a blackmail letter from her first husband.

3 ...... Mr Hebron's photograph was destroyed in a fire.

...... The photograph was not destroyed in the fire. She does not want his picture because she hated him.

...... She didn't want to show the photograph because Mr Hebron was of African descent.

4 ...... The first husband tells her to send a hundred pounds, or they will tell Mr Munro everything.

...... Effie sent the servant a hundred pounds to bring her daughter to England.

...... Effie wanted the one hundred pounds for a dress.

5 ...... Effie went out in the middle of the night because she wanted some fresh air.

...... Effie went out in the middle of the night because she wanted to see her daughter.

...... Effie went out in the middle of the night because she wanted to convince her blackmailers to leave her alone.

6 ...... Effie's first husband is in the house.

...... Effie had never seen the people in the cottage before.

...... Effie's little girl is in the house.

7 ...... Mr Munro is not sure the yellow face is a man's face.

...... Effie's first husband has a horrible yellow face because he got a terrible disease.

...... The little girl wore the mask so people would not talk about a little black girl.

PET ③ Henry and Arthur

Look at the statements below about Arthur Conan Doyle and Henry Highland Garnet. Then read the text below to decide if each statement is correct or incorrect. If it is correct, mark A. If it is incorrect, mark B.

A   B

1 'The Yellow Face' reflected the general opinion in the period when it first appeared. ☐ ☐

2 Conan Doyle had once had a very negative opinion of Africans and African Americans. ☐ ☐

3 Conan Doyle changed his opinion of black people after meeting Henry Highland Garnet. ☐ ☐

4 Henry had not been born a free man. ☐ ☐

5 Henry never had the chance to study. ☐ ☐

6 Henry fought for the liberation of slaves in America. ☐ ☐

7 Henry did not think African Americans could end slavery alone. ☐ ☐

8 Henry was against the Civil War. ☐ ☐

9 Henry changed Conan Doyle's opinions with intelligent ideas. ☐ ☐

10 'The Yellow Face' is the only Sherlock Holmes story about African Americans. ☐ ☐

Arthur Conan Doyle showed courage and sensitivity in writing 'The Yellow Face'. In 1894, when the story first appeared, many people among Conan Doyle's readers considered white people superior to black people. Also, the possibility of marriage between people with different coloured skins could not even be mentioned. But Doyle had not always held the ideas he has in 'The Yellow Face'. Earlier he had written, 'Many things have been said about the good qualities of black people. Well, my own experience has been much different: when you first meet black people you hate them, but when you know them better you hate them even more.'

Then in 1881 Doyle travelled on the Mayumba, a ship sailing to West Africa. Also travelling on this ship was an old African American man named Henry Highland Garnet. He was dying, and his last desire was to

spend some time in Africa before he died. Garnet had been born a slave 67 years earlier in Maryland. His family had escaped to Pennsylvania, which was then a state without slavery. With much difficulty Henry got a good education. He became a great speaker, writer and fighter for the abolition of slavery. He believed that the three million slaves in America could free themselves if they had the courage to do so. When the American Civil War (1861-1865) began, Henry helped to create black army units. Conan Doyle was very impressed by this modern knight who had fought for his people. He wrote about Henry, 'This gentleman did me good — a man's brain is an organ that forms thoughts and that digests the thoughts of others. Our brains always need more food!'

'The Yellow Face' is one of the two Sherlock Holmes stories that Conan Doyle wrote reflecting his new ideas about Africans and African Americans. It is interesting to note that this is also one of the few stories in which Holmes is wrong, totally wrong. Perhaps this reflects Conan Doyle's own mistake with regard to millions of people.

4 Speaking

With your partner discuss the relations between different ethnic groups in your country today. Then present your ideas to the class.

1   What are the different ethnic groups in your country?
2   Do different ethnic origins cause problems? Why?
3   Is marriage between different ethnic groups usual of unusual?

Sherlock Holmes
on Stage and Screen

In 1970, the public was promised that, 'What they didn't know about Sherlock Holmes has made a great film.' The film was *The Private Life of Sherlock Holmes*. Then, fifteen years later, filmgoers got the chance to see what Holmes was like as a boy in *Young Sherlock Holmes*. And still later Hollywood continued to offer new versions of the great detective and to reveal different details of his life. But this is not surprising: Sherlock Holmes has always been seen as a real person, and of course, people have always wanted to see what he looked like when he walked and talked.

Holmes was first filmed when his creator was still alive. Conan Doyle was always very interested in the illusions created by film. In fact, he appeared in and helped with the creation of a filmed version of his popular book about dinosaurs and ape-men, *The Lost World*. This film made history with some of the first great special effects of prehistoric monsters.

Conan Doyle was also very happy with a popular stage version of Holmes acted and written by the American William Hooker Gillette (1853-1937). Gillette modelled the look of his stage Holmes on the famous illustrations by Sidney Paget [1] (1860-1908). His play was a huge success. Conan Doyle himself said that after seeing Gillette's Holmes the character looked bad on the printed page.

Gillette did his play for 30 years. He also did a silent film version of Sherlock Holmes in 1916. This, however, was not the first time that

---

1. **Sidney Paget** : the pictures on page 82 are by this artist.

A scene from the 1970 film *The Private Life of Sherlock Holmes* (directed by Billy Wilder).

Holmes was filmed. The first was in 1900, just a few years after the invention of film projectors.

Of course, Sherlock Holmes had all the qualities to make a film star when commercial films started talking in 1927. The most famous of the early Holmes films were *The Hound of the Baskervilles* and *The Adventures of Sherlock Holmes,* which came out in 1939. Holmes was portrayed by Basil Rathbone (1892-1967) and Dr Watson by Nigel Bruce (1895-1953). These first two films were set in Victorian times just like the books. They were extremely successful, and these two actors went on to make twelve more. Many people still think that Rathbone's Holmes and Bruce's Watson are the best of all times. This is interesting when we consider that their Holmes films from 1942 on all took place in contemporary times. The film company justified this by saying that Holmes was a 'timeless character'. They even had Holmes and Watson fight against Nazi spies – World War II was going on at the time, after all.

Basil Rathbone, Ida Lupino and Nigel Bruce in the 1939 film
The Adventures of Sherlock Holmes.

After Rathbone, Holmes continued to find actors, and with the arrival of television in the 1950s, he had another medium. In fact, some Holmes fans think that the best and most accurate version of Holmes was by Jeremy Brett (1933-1995) in a British television production from 1984-1994. These versions followed very closely Conan Doyle's original stories.

But that is not all. Sherlock Holmes continues to attract television and film producers. In fact, he is the most portrayed character in the history of cinema with about 70 actors portraying him in more than 200 films (right behind Holmes in this list is another fictional character from Victorian times, Dracula).

And his fascination and charm continue to this day. Already, during the summer of 2008, two of the major Hollywood studios announced the production of two new films. One would be a comedy and the other would make Holmes into a kind of action hero and would concentrate on his fighting skills (which Conan Doyle mentioned in the original stories). This should surprise no one. After all, William Gillette's highly successful play showed Holmes fall in love. And as every fan knows, Mr Sherlock Holmes was a kind of thinking machine and had no time for women. So, why not Holmes as a comic book hero like Batman or Spiderman? After all, Holmes has already been the hero of video games for more than 20 years.

1 Comprehension check

Say whether the following statements are true (T) or false (F), and then correct the false ones.

T F

1 Major films based on the Sherlock Holmes character have usually followed the original stories by Conan Doyle. ☐ ☐

2 Conan Doyle was never very happy with the stage and film versions of his books. ☐ ☐

3 Basil Rathbone is often considered the best Holmes on film. ☐ ☐

4 Basil Rathbone's Holmes appeared as a character of the modern world. ☐ ☐

5 Jeremy Brett's Holmes is admired for its quality and accuracy. ☐ ☐

6 There have been more film versions of Dracula and Holmes than of any other fictional character. ☐ ☐

7 Today, the most important film studios seem to have lost interest in filming Holmes. ☐ ☐

8 Holmes has also become famous in the world of computer games. ☐ ☐

1 Picture summary

Look at the pictures from the three stories. Write the correct title and a suitable caption under each one.

2 Below are brief descriptions of three people. Decide which of the three stories in this volume would be the most suitable for each person.

1 Jo is interested in America. He wants to visit when he has enough money but for now he likes reading about anything that mentions America, especially its cultural aspects.

2 Tom is very romantic and has lots of girlfriends. He lives in a small town and so finds it difficult to stop them from being jealous. He is always looking for new ways to hide from them.

3 Veronica is eighteen and lives in a big city. She loves jewels but doesn't have any money to buy them.

A 'The Blue Carbuncle'

B 'The Yellow Face'

C 'A Case of Identity'

3 Book report

Write a 'book report' about the story 'The Blue Carbuncle'.

REPORT

Title		1 ....
Author		2 ....
Characters	main	3..
	minor	4 ....
Setting	place(s)	5..
	time	6 ....

Short summary

This story is about 7 ....

....

....

....

....

 **4** Summary

Read the text below and choose the correct word for each space.

'A Case of Identity'

Miss Mary Sutherland (**0**) ...A... to visit Sherlock Holmes to tell
(**1**) .... about the disappearance of her (**2**) .... Hosmer Angel. She
tells him that her father had died and then her mother remarried a
very young man, Mr Windibank. Mr Windibank sold her father's
(**3**) .... and Miss Sutherland gave the interest from her inheritance
(**4**) .... her mother, while she earned money typing.

Mr Windibank kept tight control of Miss Sutherland but she
(**5**) .... Hosmer Angel at a ball when her stepfather was away in
France for business. The young lovers could only see each other when
Mr Windibank went away (**6**) .... they wrote love letters (**7**) ....
Miss Sutherland wrote hers by hand but Mr Angel wrote his with a
typewriter.

Mr Hosmer was a little strange, (**8**) .... wearing dark glasses and
whispering, but he was gentle and kind. Miss Sutherland's mother
liked him too. Soon they decided that Miss Sutherland should marry
Mr Angel...

0	(A) comes	B becomes	C finishes	D begins
1	A them	B me	C him	D you
2	A granny	B fiancé	C brother-in-law	D father
3	A job	B money	C profession	D business
4	A from	B to	C at	D in
5	A met	B meets	C is meeting	D will meet
6	A because	B even	C but	D until
7	A yearly	B weekly	C monthly	D daily
8	A always	B never	C rarely	D hardly

 **5** Writing

Can you finish the summary above for 'A Case of Identity'?

6 The Yellow Face

Answer the following questions about 'The Yellow Face'.

1 What did the man who wanted to see Holmes leave on the table at Holmes's house?

2 How did Holmes know what the man was called?

3 Why did the man come to see Holmes?

4 What happened to his wife before he met her?

5 What did Munro's wife ask him for one day?

6 What did Munro see in the window of the cottage?

7 When Munro wanted to enter the cottage what did his wife say?

8 Who did Holmes think was blackmailing Munro's wife?

9 Who was the person with the yellow face?

10 Why did Munro's wife not tell him the truth about America when they met?

11 What did Munro decide to do about the new situation?

7 Speaking

Which of the three stories did you prefer? Explain why.

All three stories are true.

Answer to ex. 5, Page 70.

INTERNET PROJECT

Sherlock Holmes is one of the world's most famous detectives. People all over the world go to London to visit his museum at 221b Baker Street. We can visit it too without having to go to London! Ask your teacher to help you find it on the Internet and you can even have a virtual tour of Sherlock Holmes's study.

Write a short report about your visit to the museum including the following things:

▶ How the building became a museum.

▶ Describe Holmes's study: the furniture, style, colours and objects.

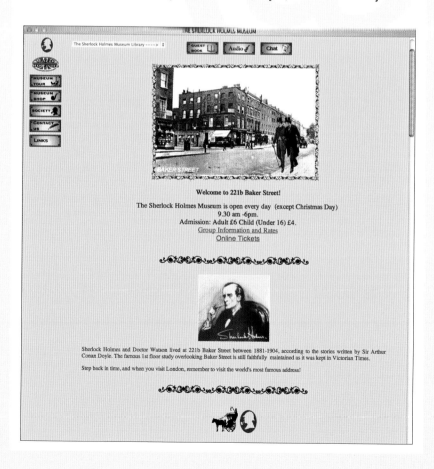

127

This reader uses the **EXPANSIVE READING** approach, where the text becomes a springboard to improve language skills and to explore historical background, cultural connections and other topics suggested by the text.

The new structures introduced in this step of our READING & TRAINING series are listed below. Naturally, structures from lower steps are included too.

The vocabulary used at each step is carefully checked against vocabulary lists used for internationally recognised examinations.

Step **Three B1.2**

All the structures used in the previous levels, plus the following:

Verb tenses
Present Perfect Simple: unfinished past with *for* or *since* (duration form)
Past Perfect Simple: narrative

Verb forms and patterns
Regular verbs and all irregular verbs in current English
Causative: *have / get* + object + past participle
Reported questions and orders with *ask* and *tell*

Modal verbs
Would: hypothesis
Would rather: preference
Should (present and future reference): moral obligation
Ought to (present and future reference): moral obligation
Used to: past habits and states

Types of clause
2nd Conditional: *if* + past, *would(n't)*
Zero, 1st and 2nd conditionals with *unless*
Non-defining relative clauses with *who* and *where*
Clauses of result: *so*; *so ... that*; *such ... that*
Clauses of concession: *although*, *though*

Other
Comparison: *(not) as / so ... as*; *(not) ... enough to*; *too ... to*

Available: